*Quantum chemistry of
atoms and molecules*

Quantum chemistry of atoms and molecules

Philip S.C. Matthews

Lecturer in Science Education
Trinity College, Dublin

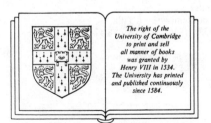

The right of the
University of Cambridge
to print and sell
all manner of books
was granted by
Henry VIII in 1534.
The University has printed
and published continuously
since 1584.

Cambridge University Press

CAMBRIDGE

LONDON   NEW   YORK   NEW ROCHELLE

MELBOURNE   SYDNEY

Published by the Press Syndicate of the University of Cambridge
The Pitt Building, Trumpington Street, Cambridge CB2 1RP
32 East 57th Street, New York, NY 10022, USA
10 Stamford Road, Oakleigh, Melbourne 3166, Australia

First published 1986

Printed in Great Britain at the University Press, Cambridge

British Library cataloguing in publication data

Matthews, P.S.C.
Quantum chemistry of atoms and molecules.
1. Quantum chemistry
I. Title
541.2′8   QD462

Library of Congress cataloguing in publication data

Matthews, P.S.C.
Quantum chemistry of atoms and molecules.
Bibliography: p.
Includes index.
1. Quantum chemistry.   I. Title.
QD462.M37   1986   541.2′8   85-17419

ISBN 0 521 24854 X   hard covers
ISBN 0 521 27025 1   paperback

OCLC 12341985

TM

Contents

Contents

Preface

Many of the concepts that are used in modern chemistry have their origins in the development of quantum theory during the first thirty years of this century. However, as with any discipline that becomes firmly established, there is a danger that the principles that lie behind the language it uses become taken for granted. It is too easy to become uncritical. The problem is compounded because quantum theory is essentially mathematical in nature. Without some understanding of the mathematics, the student is likely to miss much of the subtlety of the ideas, such as that of an atomic orbital, that he or she is expected to use.

In the pages that follow I have attempted to provide the student with the essential mathematics, and its physical interpretation, which quantum theory uses when it is applied to chemistry. Much of the mathematics is integrated into the text, but the more intricate portions are to be found in separate 'boxes'. These can be left on a first reading and returned to later. There are a number of questions at the end of nearly every section in each chapter. The questions are designed to test the student's understanding of the text and, I hope, provide fresh insights into the work. Answers are provided for all of the questions.

I have tried to give an emphasis to the fundamental ideas as they relate to bonding and spectroscopy, rather than to attempt a comprehensive treatment of quantum chemistry. Indeed, when dealing with such a complex area much has to be omitted from a text if it is to be kept to a sensible size. I have included elementary ideas on the use of symmetry but, of the omissions, group theory may be the most noticeable.

I would like to thank Drs L.L. Boyle and B.T. Sutcliffe for reading the typescript, spotting errors, giving advice and providing encouragement. At Cambridge University Press, Dr E. Kirkwood and Mrs M. Storey have been most helpful and understanding in the way they have dealt with an often tardy author.

A final word of thanks must go to my wife Margaret for maintaining the equilibrium of a house under a deluge of paper. Both Alastair and Euan are too young to appreciate the oddity that out of chaos can come isolated areas of order.

P.S.C.M.
Kedington,
May, 1985

Part 1
The fundamentals

Particles and waves

1.1 Introduction

When we first wonder about how electrons, atoms and molecules behave there is a natural inclination to think in terms of pictures or models that are based on the world about us. For the physicist, atoms in a gas can be thought to behave in many ways like a swarm of miniature billiard balls. For the chemist there is a temptation to think of an electron in an atom as a small particle revolving round the nucleus very much like a planet around the sun. Pictorial models like these help us to build mathematical models such as the kinetic theory of gases, but models invariably have their faults. In the following pages we shall need to make use of models, be they diagrams or more abstract mathematical ideas, but it is important to realise that when we deal with the submicroscopic world of electrons and atoms we must not take too much for granted.

One of the principal messages of modern quantum theory is that models based on ideas formed from observing the properties of lumps of large-scale matter (e.g. billiard balls and the like) are almost always at fault when dealing with matter of an atomic size.

1.2 Energy changes

Suppose a ball on the end of a string is given some energy sufficient to make it move in a circle. It is common experience that the ball can be made to move increasingly faster. Indeed, if we ignore relativity, apart from the limitation of the strength of the

Fig. 1.1. An energy diagram for a classical system consists of a single band.

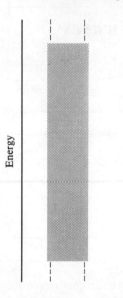

Energy

string there is no limit to the speed of rotation. Also, we could change the speed of rotation and hence the energy of the system by infinitesimally small amounts. For this system energy changes are *continuous*. This can be shown on an energy diagram which consists of one continuous band (fig. 1.1).

Energy diagrams like this occur for a wide range of systems. These are called *classical* systems. Similarly, the type of explanation that is used to describe such a system will be called a classical explanation. Classical systems together with classical explanations make up the content of classical physics. In spite of the fact that classical physics meets with great success when dealing with matter on the large scale, towards the end of the nineteenth century it became increasingly obvious that it could not satisfactorily explain the behaviour of systems of an atomic size. In particular it was realised that the energies of electrons in atoms could not change in a continuous way from one value to another. A modern example of this comes from experiments in which molecules are bombarded by X-rays. If the X-rays have the right energy, electrons are knocked out of the molecules and their energies can be measured. The difference between the energies of the ejected electrons and the incoming X-rays is a measure of the energy needed to knock the electrons out of the molecules. For propanone there are just three main values of the energy needed to eject electrons, so the resulting energy diagram has only three lines or levels (fig. 1.2).

We shall not attempt to explain the results of this type of experiment until later (section 1.4). Of more importance at present is that experiments which lead to energy level diagrams have proved impossible to explain using classical physics. If a system possesses energy levels rather than a continuous band then its energy is said to be *quantised*. Amongst other things, quantum theory attempts to explain why for systems of an atomic size energy is almost invariably quantised. In order to make a start on explaining the differences between classical and quantum theory we shall turn to the work of Max Planck.

Fig. 1.2. Energy level diagram showing the energies needed to eject electrons from propanone. The experiment is known as photoelectron spectroscopy.

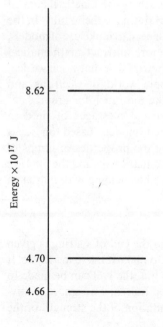

Energy $\times 10^{17}$ J

8.62

4.70

4.66

1.3 Planck's equation

If radiation is trapped in a hollow container, the radiation will give up some of its energy to the walls of the container. Similarly, the walls will begin to radiate energy back. Eventually a state of equilibrium is reached when there is no net change in the density of radiation in the container. Equilibrium radiation like this is known as black body radiation. According to classical physics, as the temperature of the container increases the energy of black body radiation becomes increasingly concentrated in the ultraviolet region of the spectrum i.e. at high frequencies. Unfortunately this is in direct conflict with experiment which shows that in fact very little of the energy is to be found at high frequencies. The failure of classical physics to account for this behaviour became known as the ultraviolet catastrophe.

In 1900 Max Planck was able to reproduce the experimental results from a theory which relied on making a sharp break with one of the main assumptions of classical physics (fig. 1.3). He denied that it was correct to believe that energy was continuously variable. Instead he proposed that for a given frequency, f, the minimum energy available was given by

$$E = hf.$$

This equation is known as Planck's equation. The constant h is Planck's constant. Its modern value is 6.626×10^{-34} Js.

In 1905 the quantity hf was called by Einstein a *quantum* of energy. The energy was also assumed to be able to change by a whole number of the basic quantum hf. Then we could have $E_1 = hf$, $E_2 = 2hf$, $E_3 = 3hf$ etc. so in general

$$E_n = nhf$$

where n is an integer. This relation embodies the quantisation of energy and enables us to draw up the type of energy level diagram which is foreign to classical physics (fig. 1.4).

Planck's constant is a measure of the *action* of a system. We hear very little of action except in advanced applied mathematics yet it is of very great interest. Imagine a ball thrown into the air; its path is well known to be a parabola. What is not immediately obvious is the property the parabolic path has that makes it more favoured than any other. The Irish mathematician Hamilton showed that the parabolic path was the one which had the minimum action (fig. 1.5).

In classical physics changes in action can be

Fig. 1.3. A schematic diagram showing the ultraviolet catastrophe for the classical solution for the energy density in a black body at some temperature, T.

Fig. 1.4. If the energy of a system is quantised according to Planck's equation $E = nhf$ then the energy level diagram consists of a set of equally spaced lines a distance hf apart.

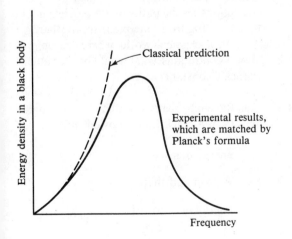

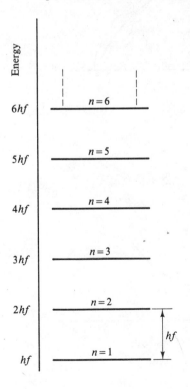

continuous. Planck was saying that changes in action could only take place in multiples of h. The fact that action is quantised is of profound importance in quantum theory and leads to some strange results when viewed from the standpoint of classical physics. On the other hand it provided Niels Bohr with a way of explaining the behaviour of simple atoms such as hydrogen. The reason for this is that action and angular momentum are closely related. If a mass, m, moves with speed v in a circle of radius r, its angular momentum is $L = mvr$. It is a simple matter to show that action and angular momentum have the same units. Whenever angular momentum appears in quantum theory, Planck's

constant is sure to be found. We shall see that angular momentum is proportional to $h/2\pi$. The quantity $h/2\pi$ has the habit of appearing in many parts of quantum theory. Because of this it has been given a special symbol, $\hbar$ (h-bar), i.e.

$$\hbar \equiv h/2\pi$$

$\hbar$ is the fundamental unit of angular momentum.

Questions

1.1 The answers to these two short calculations should give you a 'feel' for the differences in scale between the behaviour of large-scale, classical systems and their atomic counter parts.

(i) A rough estimate of the frequency with which electrons oscillate in atoms is 10^{15} Hz. If an electron has an energy 2×10^{-17} J what value does this give for n in Planck's equation? Remember that n must be an integer and that you will only obtain an estimate; not a precise value.

(ii) Now imagine a 1 kg mass to be moving at $6\,\text{ms}^{-1}$ in a circular path with a frequency of 1 Hz. What is its kinetic energy? What value of n would be needed if Planck's equation were used for this classical system?

Fig. 1.5. The relationship between action and the path of a projectile: (*a*) the actual path of a ball thrown from P to Q is a parabola; (*b*) the action for the actual path is a minimum, A_{min}. For other paths the action is greater than this value. In classical physics the graph of action varying with path is a continuous curve.

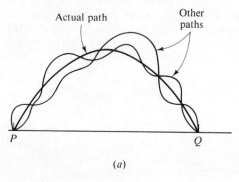

Actual path   Other paths

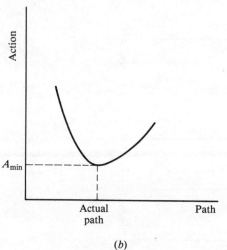

(*a*)

(*b*)

1.2 The claim that action is quantised leads to some strange results: strange that is as far as classical physics is concerned. Look again at fig. 1.5. Try to draw a similar diagram of action against path assuming that action is *not* a continuous quantity. Does this lead to any restrictions on the paths that a particle might take in going from one place to another?

In a classical world where action was always continuous, what would be the value of Planck's constant?

1.3 The Rayleigh–Jeans formula for the energy density in a black body showed that

$$\text{energy density} \propto \frac{8f^2kT}{c^3}.$$

Planck proposed that

$$\text{energy density} \propto \left(\frac{8hf^3}{c^3}\right)\frac{1}{e^{hf/kT} - 1}.$$

Here f is the frequency, k is Boltzmann's constant, c the speed of light, and T the absolute temperature.

Show that as $h \to 0$, which allows us to put $e^{hf/kT} \to 1 + hf/kT$, Planck's formula reduces to that of Rayleigh and Jeans.

1.4   What are the dimensions of action? Check that angular momentum has the same dimensions.

1.4   The photoelectric effect

In 1887 Hertz found that when light was shone on clean metal surfaces, electrons were emitted. When the values of the kinetic energy of the ejected electrons were plotted against the frequency of the light, graphs like that shown in fig. 1.6 were obtained.

There were two results from the experiments that were particularly difficult to explain.

   (i) The graphs showed that for each metal surface used the light had to have a minimum frequency before the electrons were ejected.

   (ii) The maximum energy of the electrons was independent of the intensity of the light.

The second result was at odds with classical

Fig. 1.6. Typical graph obtained in the photoelectric effect showing that the light has to have a minimum frequency, f_0, before electrons are ejected.

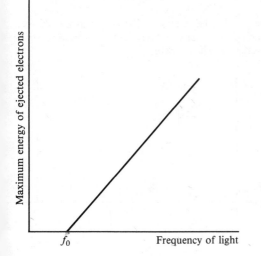

physics because, according to the accepted view of the nature of light, it was expected that if the intensity of a beam of light were increased, the energy which the beam could impart to the ejected electrons should increase. Similarly, if the intensity of the light were decreased the energy of the electrons was expected to decrease. Unfortunately for classical physics it was found that even if beams of extremely low intensity light were used the ejected electrons still had the same maximum energy as if a high intensity beam had been used. The only difference was that with low intensity beams far fewer electrons were ejected.

In 1905 Einstein applied Planck's quantum ideas to help him explain the photoelectric effect. He suggested that the energy of a beam of light was not spread evenly over a wavefront as classical physics assumed. Instead, the energy was transported in the form of 'packets' or 'bundles' of energy which were later called *photons*. According to Einstein, a beam of light of frequency f was composed of a number of identical photons all of which had the same quantum of energy given by Planck's equation, $E = hf$. An intense beam was one in which there was a large number of photons; a less intense beam simply had a smaller number of photons.

If we apply Planck's equation, photons with the minimum frequency, f_0, needed to eject electrons, would have energy hf_0. This minimum energy represents the amount of work that has to be done to remove electrons. For this reason it is known as the *work function* for the metal being used. Usually it is given the symbol ϕ. If the photons have an energy greater than ϕ the excess energy goes into increasing the kinetic energy, $\mathcal{T}$, of the electrons. Therefore, for photons of energy hf,

$$hf = \phi + \mathcal{T}.$$

Alternatively,

$$\mathcal{T} = hf - \phi.$$

We can see that this fits the graph of fig. 1.6 because if $\mathcal{T}$ is plotted against f, a straight line should be obtained whose slope is h and whose intercept on the energy axis is $-\phi$. Graphs such as this allow estimates of the magnitude of Planck's constant to be made.

To sum up, the importance of Einstein's work was that it not only provided additional evidence for the validity of Planck's ideas, but also it emphasised the fact that classical ideas about the nature of light and electromagnetic radiation in general were incomplete. In particular, Einstein had proposed that the energy of electromagnetic radiation was quantised.

Questions

1.5 The work function for sodium is approximately 3.68×10^{-19} J. If just one photon of ultraviolet light of frequency 10^{15} Hz strikes the surface of sodium (and does not merely bounce off) what would be the maximum energy of an ejected electron? What would its speed be? How would your answer compare if 10^6 photons of the same frequency hit the sodium?

1.6 The following table shows results obtained when photons of different frequencies were incident on a sample of copper.

Frequency of photon $\times$ 10^{-16} Hz	1.0	1.5	2.0	2.5	3.0	3.5	4.0	4.5
Kinetic energy electrons $\times$ 10^{17} J	0.586	0.917	1.25	1.58	1.91	2.32	2.57	2.90

Plot a graph of these results and use it to estimate the value of Planck's constant and the work function of copper.

1.5 The hydrogen atom

If hydrogen atoms are given sufficient energy, for example by exciting them electrically, then they emit electromagnetic radiation. The resulting spectrum consists of a large number of lines (fig. 1.7). There is an obvious difference between a line spectrum and a continuous spectrum such as is obtained when white light passes through a prism, or through rain drops to give a rainbow.

Of special interest is the series which appears in the visible part of the spectrum. Some seventeen years after the series was first discovered, Johannes Balmer managed to work out the formula which fits the wavelengths, λ, of the lines:

$$\frac{1}{\lambda} = R_H \left(\frac{1}{2^2} - \frac{1}{m^2} \right).$$

Here R_H is the Rydberg constant, value 1.097×10^7 m^{-1}. All the lines in the spectrum can be fitted to the formula

$$\frac{1}{\lambda} = R_H \left(\frac{1}{n^2} - \frac{1}{m^2} \right)$$

in which n, and m are integers (table 1.1). Attempts to explain the appearance of line spectra by using classical physics failed for, as we have seen, in classical physics continuous changes are expected and therefore continuous spectra should be the norm.

In 1913 Niels Bohr proposed that Planck's and Einstein's novel ideas about quantisation of

Fig. 1.7. The spectrum of hydrogen atoms shows a bewildering array of lines. Only lines in the Balmer series are shown here.

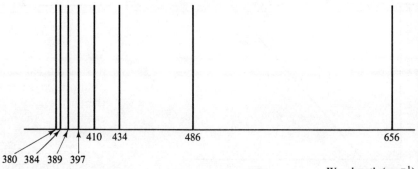

Wavelength (nm^{-1})

Table 1.1. *Series of lines in the hydrogen spectrum*

Series	Region of Spectrum	Value of n	Values of m
Lyman (1914)	ultraviolet	1	2, 3, 4, ...
Balmer (1885)	visible	2	3, 4, 5, ...
Paschen (1908)	infrared	3	4, 5, 6, ...
Brackett (1922)	infrared	4	5, 6, 7, ...
Pfund (1924)	infrared	5	6, 7, 8, ...

energy could be used to explain the line spectrum of hydrogen. His picture of the hydrogen atom was one in which the electron was a negatively charged particle constrained to move in a circular path owing to the centripetal force between it and the positively charged nucleus. We shall not follow Bohr's method here, but note that he said that the angular momentum of the electron was quantised according to the rule

$$L = n\hbar.$$

Using this condition it is possible to derive Bohr's formula for the energy levels of the hydrogen atom:

$$E_n = -\frac{e^4 m_e}{8\varepsilon_0^2 h^2 n^2}.$$

The integer n is a *quantum number* whose value determines the energy. In the formula $-e$ is the charge on the electron and m_e its mass. ε_0 is the permittivity of free space. Using the formula it is possible to display the energy level diagram for the hydrogen atom (fig. 1.8).

The energy values are negative in sign for the following reason. The energy of a completely separate proton and electron is taken as zero. As they come together the energy of the system is lowered below zero because they attract one another – we would have to put energy *in* to separate them and bring the energy back to zero. The spacing of the levels becomes increasingly close as the value of the quantum number increases. Eventually, at $n = \infty$, the energy becomes zero. This corresponds to the electron and proton no longer exerting any attraction on each other. Put in another way, when $n = \infty$ the atom is ionised.

Using Einstein's relation for the energy of a photon, if an electron drops from a level E_m to a lower level E_n, a photon of frequency f may be emitted where

$$hf = E_m - E_n.$$

Because the photon travels at the speed of light, c, its wavelength can be found from $c = f\lambda$. Thus,

$$\frac{1}{\lambda} = \frac{1}{hc}(E_m - E_n)$$

or

$$\frac{1}{\lambda} = \frac{e^4 m_e}{8\varepsilon_0^2 h^3 c}\left(\frac{1}{n^2} - \frac{1}{m^2}\right).$$

This is precisely of the form that Balmer found provided we identify the Rydberg constant as

$$R_H = \frac{e^4 m_e}{8\varepsilon_0^2 h^3 c}$$

from which its value is calculated to be nearly

Fig. 1.8. The energy level diagram of the hydrogen atom shows the levels becoming increasingly close as the quantum number, n, increases. At $n = \infty$ the atom is ionised and the electron is free to move without the influence of the proton. A continuous band results because the energy of the electron is no longer quantised.

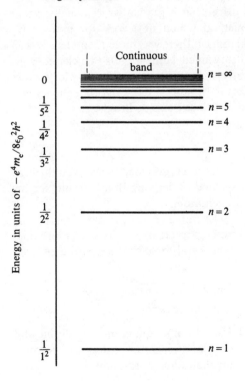

$1.0967 \times 10^7 \, \text{m}^{-1}$. This is in excellent agreement with the observed value.

Using his formula for the radius of the orbit of the electron,

$$r_n = \frac{\varepsilon_0 h^2 n^2}{\pi e^2 m_e}$$

he was able to calculate the radius for the electron in the lowest energy level, otherwise called the ground state. This radius is known as the first Bohr radius, symbol a_0. Its value 5.292×10^{-11} m is often used as a unit of length in quantum theory.

In spite of the fact that both Planck and Bohr had achieved notable successes by making assumptions about the quantisation of energy and angular momentum, neither had made a complete break with classical physics. Planck, for example, maintained for many years that his approach was merely a stop-gap measure until the 'correct' method was found to apply the ideas present in classical physics. For him quantum ideas were a necessary but temporary evil. Bohr's analysis of the hydrogen atom problem was stunning in its ability to reproduce from theory the results of experiments. Indeed the accuracy of his predictions was correct to many decimal places, but it gradually became clear that Bohr's method would not work for more complicated atoms. If there was more than one electron Bohr's theory failed. None the less Planck and Bohr were clearly on the right track. The way forward in attempting to produce a more rigorous theory was provided by Prince Louis de Broglie, to whose ideas we now turn.

Questions

1.7 This question is designed to lead you through the method of deriving Bohr's results for the hydrogen atom.

The force between the proton and electron separated by the distance, r, provides the centripetal force that keeps the electron moving in a circle i.e.

$$\frac{m_e v^2}{r} = \frac{e^2}{4\pi\varepsilon_0 r^2}. \tag{A}$$

(i) Use Bohr's quantum condition for angular momentum to derive his formula for the radius of rotation.

Now we need to work out the total energy of the electron. This is the sum of its kinetic energy, $\frac{1}{2}m_e v^2$, and its potential energy,

$$E = \tfrac{1}{2}m_e v^2 - \frac{e^2}{4\pi\varepsilon_0 r}.$$

(ii) Use equation (A) to simplify this so that m_e and v no longer appear.

(iii) Then substitute the formula you obtained in (i) for the radius. You should now have the result quoted in section 1.5.

(iv) The virial theorem says that in a potential that varies as $1/r$, we should have

$$2 \times \left(\begin{array}{c}\text{average kinetic} \\ \text{energy}\end{array}\right) = -\left(\begin{array}{c}\text{potential} \\ \text{energy}\end{array}\right)$$

or $2\mathscr{T} = -V$.

Do Bohr's results obey the theorem?

1.8 The method used in question 1.7 can be used to deal with other systems that contain only one electron. Assume that an atom has an atomic number Z so that the charge on its nucleus is $+Ze$. Then

$$\frac{m_e v^2}{r} = \frac{Ze^2}{4\pi\varepsilon_0 r^2}.$$

Also, the potential energy of the electron will be $-Ze^2/4\pi\varepsilon_0 r$. Now work out the formula for the energy levels.

1.9 Calculate the value of the energy of an electron in the lowest energy level of the hydrogen atom. This gives the energy of the ground state. What is the value of E_n when $n = \infty$?

Using your answers, calculate the ionisation energy of the hydrogen atom. How does this compare with the observed value of $1312 \, \text{kJ mol}^{-1}$? (You will need to convert to J mol^{-1} by multiplying by the Avogadro number).

1.10 By putting $Z = 2$ in your answer to question 1.8, calculate the energy needed to ionise the He^+ ion. This is the same as the second ionisation energy of helium whose value is approximately $5203 \, \text{kJ mol}^{-1}$.

1.11 Suppose a photon of wavelength 1 nm gives up all its energy to an electron in the ground state of the hydrogen atom. What will be the speed of the electron as it leaves the atom?

1.6 Waves

Classical physics does not deal only with particles; of almost equal importance is the way it explains the properties of waves. De Broglie's contribution to the advance of quantum theory was to introduce arguments based on the behaviour of waves. In order to see the force of his contribution we shall first revise some of the properties of waves.

On the large scale it is easy to distinguish visually between waves and particles. No one, for example, would confuse a billiard ball with a wave on the sea. Waves, unlike particles, do not have a single exact position – waves are extended in space as indicated by the idea of wavelength. Neither is it found convenient to work out the energy carried by a wave by referring to its momentum; momentum is a particle-like property. Indeed, most of the mathematical methods which are used to describe wave motion in classical physics are different from those applicable to particles.

The two waves shown in fig. 1.9 have the same amplitude, A, wavelength, λ, and will be assumed to move with the same speed, v. Because $v = f\lambda$ they will have the same frequency, f. The waves will interfere completely and destructively because they are completely out of phase. One of them lags exactly half a wavelength behind the other. By comparison the two waves in fig. 1.10 are exactly in phase and interfere completely constructively.

Of course waves do not always interfere completely constructively or completely destructively because they are not always perfectly in or out of phase. More often they interfere only partially and give rise to a resultant which is quite unlike the original wave. A particularly important example of this is the formation of a *wave packet*. A wave packet is a wave which has zero intensity everywhere except in a very small region of space. It is possible to build up a wave packet by the interference of a large number of waves which have very slightly different wavelengths (or frequencies) (fig. 1.11).

Fig. 1.9. Two waves which have the same amplitude, A, and which are completely out of phase give a resultant of zero amplitude. There is complete destructive interference.

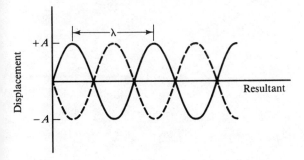

Fig. 1.10. In this case two waves of equal amplitude are completely in phase and interfere completely constructively.

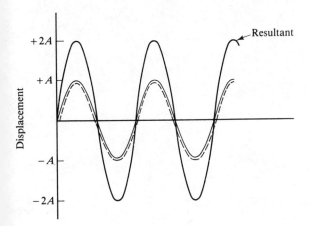

Fig. 1.11. The figure shows the outcome of a computer simulation of a wave packet built up from the interference of a large number of waves.

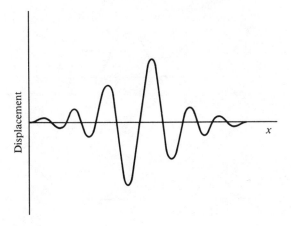

In general the more waves that are superimposed the sharper is the wave packet. However, there is one very important point about wave packets that we shall need to refer to later. That is, although a wave packet may be found localised in a small region of space at one moment, it cannot be guaranteed that it will remain localised at a later time. In brief, many, but not all wave packets, tend to spread in time (fig. 1.12).

A further property of waves is that they can be diffracted. A simple way of illustrating diffraction is to pass a monochromatic beam of light, such as can be obtained from a laser, through a narrow slit (fig. 1.13). On the further side of the screen, along AB, can be placed a photographic plate. Alternatively a photocell could be moved along AB so as to act as a detector for the light. A similar experiment can be done using two slits instead of one, in which case it is known as Young's double slit experiment. When the photographic plates from

double and single slit experiments are developed they are found to be noticeably different. The difference between them is shown up most starkly if the outputs of the photocell measurements are plotted against position along AB (fig. 1.14).

We can see that in the double slit experiment

Fig. 1.13. Experimental arrangement for a single slit diffraction experiment. In Young's experiment there would be two slits in the screen.

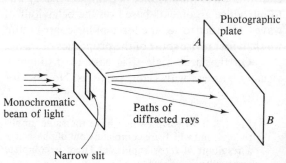

Fig. 1.12. The spreading of a wave packet as time increases: (a) the wave packet is localised at x_0 at a time, t_0; (b) at a later time t_1 the packet is at x_1 but has spread; (c) at an even later time, t_2, the packet is centred on x_2 but has spread to such an extent that it is no longer sharply defined at a particular point.

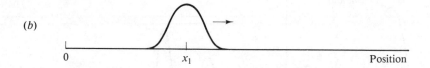

light reaches positions on the screen which are in darkness in a single slit experiment. Similarly some parts of the screen are in darkness which would be illuminated in the single slit experiment. Classical physics has an explanation for these results which was originally thought out by Huyghens. Huyghens' principle relies on using the phenomenon of interference to explain the patterns, but we shall not discuss it here. Rather we shall simply note that diffraction represents a phenomenon which is incontrovertibly associated with wave motion. The classical idea of particles excludes the notion that they can be diffracted or interfere with each other. It is worth remembering these points because we shall now consider some experimental evidence that

shows diffraction to be a more complex occurrence than the early classical physicists imagined.

In 1909 the Cambridge physicist G.I. Taylor showed that diffraction patterns could be obtained even if the source of light was so weak that the patterns took months to form. If his experiments were repeated with modern equipment so that the output of a photocell could be connected to a loudspeaker then a series of 'clicks' would be heard. This means that energy arrives at the photocell in all or nothing fashion. There is nothing gradual about the process; either a click is heard or it is not. This sort of behaviour is more like that which we normally associate with particles, and not with waves. Thus on the one hand waves do not appear to arrive at the detector; instead small packets, or *quanta*, of energy arrive which suggests that we might be dealing with particles.

Another peculiarity about diffraction shows up if, still using a very weak source of light, the photographic plate is developed after being exposed to the light emerging from the slit, or slits, for progressively longer times (fig. 1.15). A plate devel-

Fig. 1.14. Single and double slit diffraction patterns: (*a*) intensity pattern for a single slit diffraction experiment; (*b*) intensity pattern for a double slit diffraction experiment.

Fig. 1.15. These three sketches illustrate the appearance of a photographic plate in a diffraction experiment if it were developed at various times after the start of the experiment. (*a*) Appearance of plate soon after the start of the experiment. (*b*) Appearance some time later. (*c*) Appearance towards the end of the experiment.

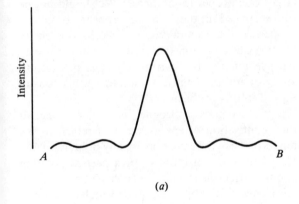

(*a*)

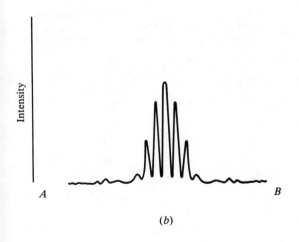

(*b*)

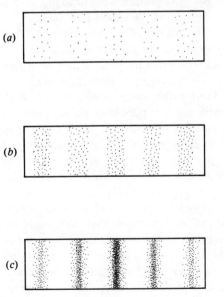

(*a*)

(*b*)

(*c*)

oped soon after the start of the experiment displays a large number of dots – a fact which supports the idea of the light arriving in quanta. What is more, the arrangement of the dots is entirely random. It is only after a long period of time when a sufficiently large number of quanta has arrived at the plate that we begin to recognise the outline of the full diffraction pattern. Thus it seems that the diffraction pattern represents a statistical average of a large number of arrivals of light quanta. If we were to attempt to predict where the next individual quantum would arrive on the plate we would meet with complete failure. It is impossible to predict with certainty where the dots will appear.

We should remember that not only visible light but all forms of electromagnetic radiation can produce diffraction patterns if a suitable grating can be found. A particularly useful example of this is the diffraction of X-rays by crystal lattices, a phenomenon which was first successfully explained by Sir Lawrence Bragg. He showed that if the spacing between a set of crystal planes is d and the wavelength of the X-rays is λ then there would be constructive interference of the diffracted rays when

$$n\lambda = 2d \sin\theta.$$

Here n is an integer and θ is the angle which the incoming rays make with the crystal planes. Looked at in another way we can see that if d is known and θ measured, it is possible to work out the value of the wavelength. Bragg's equation was put to just such a use by two Americans, Davisson and Germer, in 1927. They discovered, albeit by accident, that *electrons* incident on a crystal of nickel gave rise to a diffraction pattern. From the pattern they were able to calculate the wavelength associated with the electrons. Their results were confirmed by G.P. Thomson who actually published photographs of diffraction patterns obtained by passing electrons through thin films of gold. The production of diffraction patterns was clear evidence that in some experiments electrons showed wave-like properties.

To be set against this is the fact that some four years earlier, in an experiment that now bears his name, Compton had showed that when γ-rays encountered electrons the way they behaved obeyed the laws of conservation of momentum. This meant that in the Compton effect electrons and electro-magnetic waves were behaving in a way that we would normally associate with particles.

The net result of the Compton effect and Davisson and Germer's experiments is summed up by saying that electrons and electromagnetic radiation show a *wave–particle duality*. This apparently eccentric behaviour did not come as a surprise to de Broglie. In 1924 he had derived an equation which directly related the particle-like property of momentum to the wave-like property of wavelength:

$$mv = h/\lambda.$$

The method he used to derive the equation was exceedingly complex, employing ideas present in the old quantum theory and relativity.

De Broglie claimed that his equation was true for all matter. Hence the name *matter waves* to describe the waves. If this seems implausible we should remember that the concepts of particles and waves are used as a result of our experience of the large-scale properties of matter. We should not be too surprised if they do not appear to fit the properties of the submicroscopic world. It is one of the tasks of quantum theory to explain the wave–particle duality, but fortunately it is possible to study quantum chemistry without getting too involved in this problem.

To see how de Broglie's relation can lead to the quantisation of energy we shall return to the model of the hydrogen atom used by Bohr (fig. 1.16). He assumed the electron to be a particle rotating about the nucleus in a circle of constant radius for a given energy. If we now accept de Broglie's idea and represent the electron as a wave, this wave must fit

Fig. 1.16. Bohr's model of a hydrogen atom consisted of the electron, e, rotating about the nucleus in a path of constant radius, r, for a given energy.

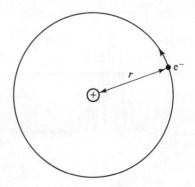

on the circumference of the circle. However, if destructive interference is to be avoided then the wavelength must fit an exact number of times on the circumference (fig. 1.17). Mathematically this means that

$$n\lambda = 2\pi r$$

where n is an integer. Using de Broglie's relation and remembering that the angular momentum, L, is given by mvr it is simple to obtain Bohr's equation for the quantisation of angular momentum

$$L = n\hbar.$$

Thus we have shown how starting from a simple description of how waves behave we can obtain equations in which integers appear quite naturally. Therefore it is not surprising that de Broglie's work leads to a new way of exploring quantum effects and explaining the appearance of quantum numbers.

A word of warning though; it would be a mistake to imagine that the wave picture of the electron in an orbit actually shows the electron moving around the nucleus in an up-and-down path. We can also scotch another possibility that can be appealing. It is tempting to think of an electron as a wave packet of minute dimensions. The problem with this idea is that, as we have seen, wave packets tend to spread in time in all but a few exceptional cases. Thus even if an electron started out as a wave packet it would soon spread out and no longer be localised.

Questions
1.12 Davisson and Germer recorded the energies of the electrons they used in electron volts, eV. If the speed of the electrons is small compared with the speed of light then relativistic correc-

tions need not be made and we can give the kinetic energy of the electrons directly in terms of eV. The number of electron volts is given by multiplying the magnitude of the electronic charge, e, by the accelerating voltage, V. Then

$$\tfrac{1}{2}m_e v^2 = eV.$$

(i) Show that this can be re-arranged to give

$$m_e v = \sqrt{(2m_e eV)}. \qquad \textbf{(B)}$$

There is a set of planes in a nickel crystal for which the spacing d appearing in the Bragg equation is 1.075×10^{-10} m. Davisson and Germer noticed a reflection at an angle 75° when using electrons accelerated through 35 V.

(ii) Multiply the voltage by the charge on the electron to work out the energy of the electrons used. Then use Bragg's equation to calculate the wavelength of the electrons. Assume that a first order reflection was observed, for which $n = 1$.

(iii) Compare your answer with the wavelength calculated from equation (B) and de Broglie's relation.

1.13 Estimate the mass and speed of a golf ball and a car. What would be the de Broglie wavelength of each of them? Why is it that we never observe diffraction patterns made by such large particles?

1.14 Hydrogen molecules are amongst the most massive particles to produce an observable diffraction pattern. The mass of a hydrogen molecule is approximately 3.4×10^{-27} kg. If

Fig. 1.17. These two waves have wavelengths that fit an exact number of times around the circumference of the circle. There is no destructive interference.

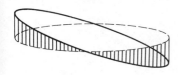
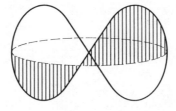

the molecules move at a speed $1700\,\mathrm{m\,s^{-1}}$, what is the associated de Broglie wavelength?

1.7 Summary

We have seen that quantum theory developed because classical physics was unable to explain why changes in action and energy were abrupt rather than continuous. Bohr was particularly successful in explaining the form of the hydrogen spectrum through applying Planck's equation $E = hf$ together with Einstein's new photon theory of electromagnetic radiation.

De Broglie showed that quantum conditions could arise out of the properties of his matter waves, but at the price of having to account for wave–particle duality expressed in his relation $mv = h/\lambda$.

The modern form of quantum theory began with the publication of the work of Erwin Schrödinger in January 1926. He derived an equation that governed the behaviour of de Broglie waves which could not only reproduce all Bohr's results but much more besides. A new theory was needed because in spite of the success of the old quantum theory in reality it had many failings. It was unable to explain the spectral and other properties of atoms with more than one electron.

Schrödinger's equation, and its solutions, form the subject of the next chapter.

2

Schrödinger's equation and wavefunctions

2.1 Introduction

In this chapter we will try to make sense of the Schrödinger equation and see how it can be solved in a few simple cases. The examples we shall use are not at first sight chemical ones; chemical examples are far too complicated for an initial study. However, they have the virtue of showing nearly all the points which we shall use in later chapters. Especially, we shall investigate Max Born's interpretation of the solutions of the Schrödinger equation. Born's ideas marked a final break with classical physics and opened the way for the modern theories of chemical bonding.

2.2 Schrödinger's equation

When a hydrogen atom is ionised, the electron has moved so far away from the proton at the nucleus that it no longer feels any attractive force pulling it back. The electron is able to move freely and its energy can change in a classical manner. This is why there is a continuous band in the hydrogen spectrum as well as the lines. We will assume that de Broglie's ideas hold, so it should be possible to describe the motion of the electron using the mathematics of waves.

We shall choose a sine wave to describe the electron (fig. 2.1). Restricting the electron to the x-direction, we shall write the wavefunction as

$$\psi = A \sin (2\pi x/\lambda).$$

The amplitude of the wave is A. Later we shall see

how the value of the amplitude can be worked out.

Because the electron is moving freely its potential energy will be zero. This means that its total energy, E, will be equal to its kinetic energy

$$E = \tfrac{1}{2}m_e v^2$$

or

$$E = \frac{1}{2m_e}(m_e v)^2.$$

Using de Broglie's relation,

$$E = \frac{1}{2m_e}\left(\frac{h}{\lambda}\right)^2$$

so that

$$\lambda = \frac{h}{\sqrt{(2m_e E)}}.$$

Fig. 2.1. A comparison of a simple sine wave and a wavefunction: (a) a simple wave shape is given by $y = \sin \theta$ where θ is measured in radians; (b) a wavefunction showing variation with position, x, which has the characteristics of a sine wave, is $\psi = A \sin(2\pi x/\lambda)$.

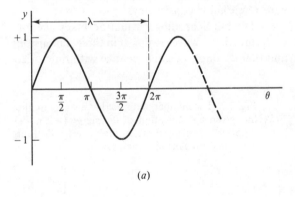

(a)

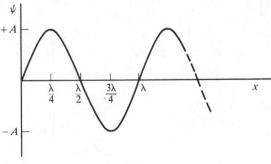

(b)

This equation shows that as the energy increases, the wavelength becomes shorter, and vice versa (fig. 2.2). The calculus tells us that the slope of the wavefunction is given by $d\psi/dx$. When the form of the wave is changing rapidly from peak to trough, the slope changes rapidly. The change in slope is $d^2\psi/dx^2$ and its value also increases as the wavelength decreases. This implies that $d^2\psi/dx^2$ and the energy are related in some way. It is not difficult to show that (M.2.1)

$$-\frac{\hbar^2}{2m_e}\frac{d^2\psi}{dx^2} = E\psi.$$

This is the simplest form of the equation which Schrödinger derived in 1926, and which marked the start of modern quantum theory. Basically it gives the recipe for finding the energy associated with a de Broglie wave. We have worked towards the equation by a route totally unlike Schrödinger's. His method was involved and mathematically rigorous. But this is not to say that he proved that his equation must be correct. The test of his equation, like that of any other in science, is to find if it gives results which agree with experiment. This it does with remarkable consistency.

The full Schrödinger equation takes account of the fact that a wave may exist in three dimensions and that its amplitude may change with time. Also,

the total energy will be the sum of a kinetic and a potential energy term. Until further notice we shall ignore problems in which time dependence is important. The solutions of Schrödinger's equation which do not change in time describe how stationary states behave. The most general form of the Schrödinger equation for stationary states is ·

$$-\frac{\hbar^2}{2m_e}\left(\frac{\partial^2\psi}{\partial x^2} + \frac{\partial^2\psi}{\partial y^2} + \frac{\partial^2\psi}{\partial z^2}\right) + V\psi = E\psi$$

or

$$-\frac{\hbar^2}{2m_e}\left(\frac{\partial^2}{\partial x^2} + \frac{\partial^2}{\partial y^2} + \frac{\partial^2}{\partial z^2}\right)\psi + V\psi = E\psi$$

We have used partial differentials like $\partial^2/\partial x^2$, rather than d^2/dx^2, to take account of the fact that ψ may depend on all three coordinates x, y and z (see M.2.1). Schrödinger's equation can also be written in a more compact form as

$$-\frac{\hbar^2}{2m_e}\nabla^2\psi + V\psi = E\psi.$$

∇^2 is the Laplacian operator. It stands for $(\partial^2/\partial x^2) + (\partial^2/\partial y^2) + (\partial^2/\partial z^2)$. We can safely leave details of this and other operators until chapter 6.

The task of quantum theory is to solve Schrödinger's equation in order to determine the allowed energy levels, and to determine the correct form of the wavefunction for each energy. Sur-

Fig. 2.2. The number of nodes that a wavefunction has given an indication of its energy ($* \equiv$ node). (a) Short wavelength means high energy. (b) Long wavelength means low energy.

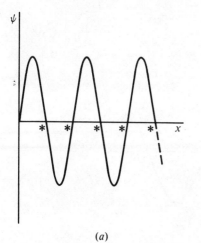

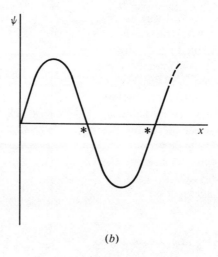

(a)   (b)

prisingly, it turns out that if we know the wavefunction then we can calculate all the properties of a system that we can measure by experiment.

2.3 The particle in a box

An electron is held in an atom by the attraction between its negative charge and the positive charge at the nucleus. The potential energy which arises from this attraction depends on $1/r$, the inverse of the distance of the electron from the nucleus. The inclusion of a term like this makes the Schrödinger equation very difficult to solve. For this reason we shall begin our study by solving a much simpler

Fig. 2.3. The particle in a one-dimensional box cannot escape over the potential energy barriers of infinite height.

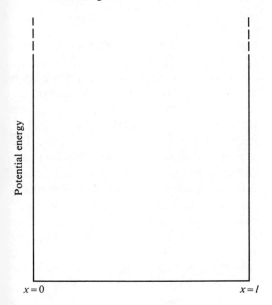

problem, but one which illustrates many of the facets of more complex cases.

The idea is that we trap a particle in a box which has infinitely high walls. The walls of the box form potential energy barriers which prevent the escape of the particle (fig. 2.3). Within the box the particle's potential energy is zero. To get out of the box its potential energy would have to become infinitely great; this is an occurrence that we can discount. To simplify matters even further we assume that the particle can only move in one dimension, along the x-axis say.

A classical particle would move to and fro across the box bouncing off the walls. In de Broglie's terms this motion would be associated with two waves, one travelling to the right and one to the left. If they are not continually to interfere destructively then standing waves must be set up. We can liken these waves to those which can be set up on a string of the same length as the box (fig. 2.4). As the amplitudes of the waves are zero at $x = 0$ we should expect sine waves to be involved. Also, the amplitudes are zero at $x = l$. Now, $\sin \theta$ is zero whenever $\theta = 0, \pi, 2\pi, 3\pi, \ldots$, i.e. when

$$\theta = n\pi.$$

Thus, whereas we had $\psi = A \sin(2\pi x/\lambda)$ for the free electron, at $x = l$ we must now have

$$\frac{2\pi l}{\lambda} = n\pi$$

or

$$\frac{2\pi}{\lambda} = \frac{n\pi}{l}$$

where n is an integer greater than 0. For $n = 0$ we would obtain a wave of infinite length, which would

Fig. 2.4. The three simplest standing waves that can be set up on a string; the sketches show the relationship between the length, l, of the string and the wavelength, λ. (a) $\lambda = 2l$; (b) $\lambda = l$; (c) $\lambda = 2l/3$.

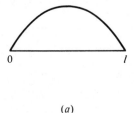 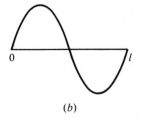 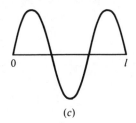

(a)   (b)   (c)

clearly not fit in the box. The result is that

$$\psi_n = A \sin(n\pi x/l).$$

Given our restrictions on the box (zero potential energy and no movement in the y- or z-directions) the Schrödinger equation is

$$-\frac{\hbar^2}{2m}\frac{d^2\psi}{dx^2} = E\psi.$$

Because (use M.2.1)

$$\frac{d^2\psi_n}{dx^2} = -\frac{n^2\pi^2}{l^2}\psi_n$$

we have

$$E_n = \frac{n^2h^2}{8ml^2}$$

Fig. 2.5. The first four energy levels and wavefunctions for a particle in a one-dimensional box

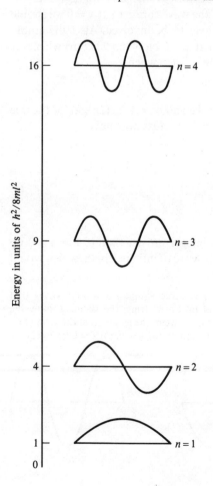

where the subscript n has been added to E to show that its value depends on the value of the integer, or quantum number, n.

The derivation did not depend on the value of the amplitude, A, of the waves, but for reasons which we shall give later (section 2.8) it is always chosen to be $\sqrt{(2/l)}$, i.e.

$$\psi_n = \sqrt{\left(\frac{2}{l}\right)}\sin\left(\frac{n\pi x}{l}\right).$$

In fig. 2.5 are shown the first four wavefunctions superimposed on the energy levels to which they belong. As we expected, the wavelength decreases as the energy increases. It should be obvious that with increasing energy, the number of nodes also increases. This is a general property of all wavefunctions.

Questions

2.1 Consider a molecule of buta-1,3-diene, CH_2=CH—CH=CH_2. If we assume that an electron belonging to one of the carbon atoms can move along the length of the molecule then we can think of the electron as trapped in a one-dimensional box equal in length to the length of the molecule. The average of a single and a double bond between carbon atoms is 0.14 nm so we could put l = 0.42 nm. However, a more realistic model is one where the electron could reach some way beyond the terminal carbon atoms; say half an average bond length at each end. Thus we shall take l = 0.56 nm.

(i) Work out the energies of the first five energy levels. Show them on an energy level diagram.

(ii) Using Planck's equation in the form $E = hc/\lambda$, calculate the wavelength of the lines in the spectrum that would be obtained if the electron could move between E_1 and E_2, E_2 and E_3, E_3 and E_4, E_4 and E_5.

(iii) Which of these comes nearest to the observed result, $\lambda \approx 220$ nm?

2.2 Follow through the method of question 1 for octa-1,3,5,7-tetraene, CH_2=CH—CH=CH—CH=CH—CH=CH_2. Which transition between energy levels fits the observed wavelength $\lambda \approx 280$ nm? Hint: write a general formula for λ for transitions between any two levels.

2.3   The particle in a one-dimensional box model of a molecule is otherwise known as the free electron molecular orbital model. In the light of your answers to questions 2.1 and 2.2, do you think the model to be a good one? Why?

2.4   Zero point energy

The formula for E_n shows that as the length of the box increases the spacings between the energy levels decrease. Eventually, when l becomes infinitely large the levels are so close that the energy changes are continuous, and we return to the behaviour expected in classical physics. On the other hand, if we attempt to confine the particle to a box of shorter and shorter length, the values of E_n increase markedly.

However, of more interest than these points is the fact that the energy of the ground state (the state with the lowest energy), E_1, is not zero. This is in direct conflict with our classical, common-sense way of thinking. We would expect the lowest energy level to be the one where $E = 0$. Quantum theory says that this is not so. Even at absolute zero the particle in a box would have the minimum energy $E_1 = h^2/8ml^2$. As a result E_1 is known as the zero point energy although, in a less emotive way, it would simply be called the ground state energy. The solution of Schrödinger's equation for many other systems produces a similar result, and there is good experimental evidence to show that it is not a fiction.

The explanation of zero point energy lies with Heisenberg's uncertainty principle, to which we will turn in chapter 6.

Questions
2.4   Calculate the zero point energy of an electron trapped in a box of length 1 nm. Compare this with the corresponding result for, say, a squash ball in a squash court. (Estimate the mass and length.)

2.5   Use your answer to question 2.4 to explain why zero point energy is foreign to classical physics.

2.5   Degeneracy

The particle in a box problem can be made more realistic if we allow the particle to move in two dimensions (fig. 2.6). Movement in the x-direction and the y-direction means that

$$-\frac{h^2}{2m}\left(\frac{\partial^2\psi}{\partial x^2}+\frac{\partial^2\psi}{\partial y^2}\right)=E\psi.$$

We shall write the solutions of the one-dimensional box as $X(x) = \sqrt{(2/l)}\sin(n_x\pi x/l)$. Perhaps surprisingly $X(x)$ remains a solution of the two-dimensional box Schrödinger equation (M.2.2). This should give us a clue to the form of the wavefunctions that depend on y. Because the box is square, the x- and y-axes are interchangeable and we write $Y(y) = \sqrt{(2/l)}\sin(n_y\pi y/l)$. Clearly, the total wavefunction must be a combination of $X(x)$ and $Y(y)$. The choice we shall make is $\psi(x,y) = X(x)Y(y)$ or

$$\psi(x,y)=\frac{2}{l}\sin\frac{n_x\pi x}{l}\sin\frac{n_y\pi y}{l}.$$

This can be shown (M.2.3) to be a solution of the Schrödinger equation, giving rise to the energy levels

$$E_{n_x,n_y}=(n_x{}^2+n_y{}^2)\frac{h^2}{8ml^2}.$$

Fig. 2.6. Potential barrier for a particle in a two-dimensional box.

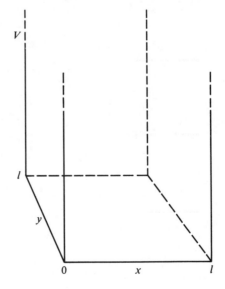

In passing we should note that our approach illustrates two common tactics in quantum theory. First, we try to take account of any symmetry in the system under study. Secondly, if the wavefunction depends on a number of variables, such as the x, y, and z coordinates, then it is sensible to attempt to find solutions which separately depend on a product of wavefunctions for each of the variables e.g. $\psi(x, y, z) = X(x)Y(y)Z(z)$.

The lowest energy level is the one for which $n_x = 1$, $n_y = 1$. The next is when $n_x = 2$, $n_y = 1$ or when $n_x = 1$, $n_y = 2$. This shows that there are two solutions $((2/l)\sin(2\pi x/l)\sin(\pi y/l)$ and $(2/l)\sin(\pi x/l)\sin(2\pi y/l))$ which have the same energy. These two solutions are degenerate solutions. A set of solutions of the Schrödinger equation whose members have identical energies is a degenerate set. Many of the solutions of the two-dimensional square box are degenerate (fig. 2.7).

This is a direct consequence of the symmetry of the box. If we had chosen a box with unequal sides the degeneracy would be removed. The hydrogen atom is a particularly symmetrical system. It will be no surprise to learn that there are many degenerate sets of solutions of the Schrödinger equation in this case.

Fig. 2.7. Energy levels of the particle in a two-dimensional box, the individual levels are labelled by the quantum numbers n_x, n_y. The total energy is proportional to $n_x^2 + n_y^2$.

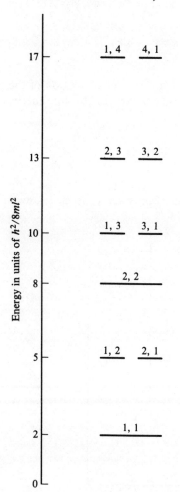

Questions

2.6 Use the formula for $\psi(x, y)$ when $n_x = n_y = 1$ in order to try to sketch contour diagrams of $\psi(x, y)$. Repeat for $n_x = 1$, $n_y = 2$ and $n_x = 2$, $n_y = 1$. If you have access to a computer it would help to write a program to calculate a series of values of $\psi(x, y)$ as x varies, keeping y constant; and vice versa.

2.7 It is possible to imagine a particle stuck in a three-dimensional cubic box. By making an analogy with the results for the one- and two-dimensional boxes
   (i) write down the Schrödinger equation,
   (ii) guess the formula for the energy, E_{n_x, n_y, n_z},
   (iii) guess the expression for the wavefunctions, $\psi(x, y, z)$.

If you have time, show that your wavefunction is indeed a solution of the Schrödinger equation with the required energy.

2.8 Draw an energy level diagram for a particle in a three-dimensional cubic box. Show on it a separate line for each degenerate solution. What is the degeneracy of the levels which have the same energy as $E_{1,2,3}$?

2.6 Combining wavefunctions

It might be thought that we have found the only correct solutions for the particle in a box. In fact there are others. In particular, let us take any pair of wavefunctions for a degenerate state, which we will call $\psi_1(x, y)$ and $\psi_2(x, y)$. If we combine them to give

$$\psi_+ = \psi_1(x, y) + \psi_2(x, y)$$

and

$$\psi_- = \psi_1(x, y) - \psi_2(x, y)$$

then both these combinations are respectable solutions having the same energy as the original functions (M.2.4).

We shall see that results like this are of immense importance in the theory of chemical bonding. It is an example of the principle of superposition. This says that if we have obtained solutions of the Schrödinger equation, provided certain conditions are met, we are free to create further solutions from them to suit our needs. One condition is that the combinations should be linear. They should not, for example, involve combining the squares or cubes of wavefunctions. In symbols, the principle says that if ψ_1 and ψ_2 are acceptable wavefunctions then the combination

$$\psi = a\psi_1 + b\psi_2$$

is also a solution of the Schrödinger equation; a and b are constants, not necessarily integers.

Questions

2.9 If you followed the working of M.2.4 for ψ_+, show that ψ_- is also a solution with energy E_1. Repeat for $\psi = a\psi_1 + b\psi_2$.

2.10 Look again at question 2.6. Repeat the method, but this time sketching ψ_+ and ψ_-. How do the two sets of sketches compare? Where are the nodes in the two sets?

2.7 Interpreting wavefunctions

Although we have met some of the properties of wavefunctions, we have not yet dealt with their true significance. In this section we shall try to explain the 'meaning' of the wavefunctions. The interpretation which has found the most favour was originally proposed by Max Born in 1926. The peculiarity in his approach was that he preferred to talk about probabilities rather than certainties when explaining atomic events. Because of the importance of his ideas we will first develop a little of the theory of probabilities.

Information which is collected from experiment or simple observation forms either a discrete

Tabe 2.1. *Statistical formulae for use with discrete and continuous distributions*

Discrete distributions	Continuous distributions
$\sum P_i = 1$	$\int P(x)\,dx = 1$
$\mu = \sum x_i P_i$	$\mu = \int x P(x)\,dx$

or a continuous distribution. An example of a discrete distribution would arise from counting the cars on a stretch of road at various times of day. The data would always be a whole number; fractions of cars are not allowed. An example of a continuous distribution would be the distances between the cars. Measurements of distance could be any positive number and could merge imperceptibly into one another.

In a discrete distribution we shall call any one of the measurements, x_i, and the probability of measuring this value P_i. The mean, μ, of the results is

$$\mu = \sum x_i P_i$$

where the summation takes place over all the individual measurements. Also, because the total probability for all measurements must be unity (certainty),

$$\sum P_i = 1.$$

An example of the use of these formulae is worked through in M.2.5.

If we attempt to use these formulae for continuous distributions there will be trouble; we would need to sum an infinite number of terms. Fortunately the calculus provides a way round the problem. If we use dx to represent the infinitesimally small difference between two measurements, the probability of a measurement within this range can be written $P(x)dx$. Here $P(x)$ is a continuous function of x. The summation of the discrete distribution is replaced by integration as shown in table 2.1.

$P(x)$ is the probability *density*, not to be confused with the probability, P_i, for a discrete distribution. We shall examine some of its properties shortly, but now we return to Max Born.

Born's apparently perverse suggestion was that the wavefunction, ψ, alone was incapable of any interpretation that would fit with experimental results. Rather, he said that we should look at the square of the magnitude of the wavefunction, $|\psi|^2$, and identify this as a probability density function. To see how this works we shall compare the classical and quantum theory approaches to the particle in a box problem.

The classical theory

The classical probability density is constant along the length of the box (M.2.6):

$$P(x) = 1/l.$$

As a consequence the graph of $P(x)$ against x is particularly simple. The integral $\int P(x)\,\mathrm{d}x$ represents the area under the graph. We know that this integral should be exactly one, and a glimpse at fig. 2.8 shows this to be the case. Also, we would expect the mean position to be at $x = l/2$. This proves to be the case because

$$\mu = \int_0^l x(1/l)\,\mathrm{d}x$$

$$= \frac{1}{l}[\tfrac{1}{2}x^2]_0^l$$

so

$$\mu = l/2.$$

The quantum approach

For the one-dimensional box the solutions of the Schrödinger equation were

$$\psi_n = \sqrt{(2/l)}\sin{(n\pi x/l)}.$$

Fig. 2.8. Graph of the classical probability density plotted against position for a particle in a one-dimensional box. In this case $P(x) = 1/l$. The area under the graph is $(1/l) \times l = 1$.

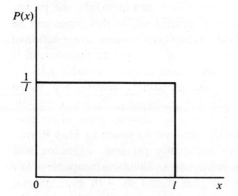

With a little effort we can prove that (M.2.7)

$$\int_0^l |\psi_n|^2\,\mathrm{d}x = 1$$

which fits with our requirement that $\int P(x)\,\mathrm{d}x = 1$. Similarly we can show that

$$\mu = \int_0^l x|\psi_n|^2\,\mathrm{d}x$$

gives

$$\mu = l/2$$

which agrees with the classical result. However, it would be a mistake to think that the quantum and classical theories agree in all respects. For example, the graphs of $|\psi_n|^2$ are very different from the classical graph (fig. 2.9). The quantum probability distributions show that, contrary to classical theory, the probability of finding the particle is not uniform along the length of the box. Especially, in the ground state (wavefunction ψ_1) the particle is much more likely to be found towards the centre of the box. As the energy increases the probability of finding the particle towards the edges increases at the expense of the initial build up of probability density at the centre. As the quantum number, n, increases the classical distribution takes on the appearance of an average of the maxima and minima in the quantum distribution.

Irrespective of the form of the graphs we should be careful to note that the probability of finding the particle at a single point is zero. This is true in both classical and quantum approaches. It follows from the fact that points have no finite dimension. Therefore, along the length of the box there will be an infinite number of points and the probability of finding the particle at any one of them will be infinitely small, i.e. zero. The moral of this is that we should be careful not to confuse probabilities with probability densities. The presence of zeros in the quantum probability density distributions means that near these points there is a very small probability of finding the particle. Near the maxima there is a much higher probability of finding the particle. In the classical case there is no corresponding fluctuation in the probabilities of finding the particle in a particular region.

Fig. 2.9. Graphs of the probability density for five states of the particle in a one-dimensional box. The graph for $n = 10$ has a different scale for the length of the box from the other four. The dashed lines show the shapes of the wavefunctions. The chain line in the graph for $n = 10$ shows the classical probability density.

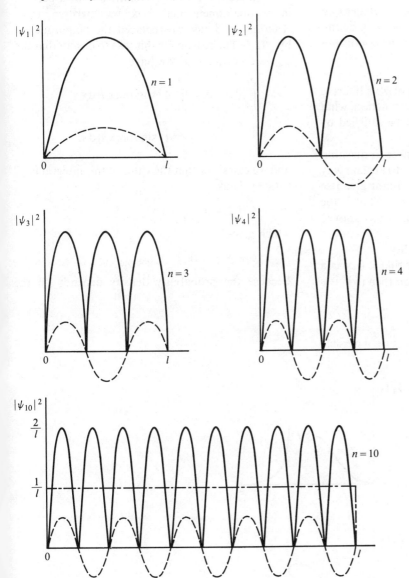

Questions

2.11 In the theory of wavefunctions the following conditions are usually laid down:
   (a) a wavefunction must have only one value at any given point;
   (b) a wavefunction must vary in a continuous fashion;
   (c) a wavefunction must be finite at all points.

   In fig. 2.10 are shown some sketches that are supposed to represent wavefunctions. Which of them do not satisfy the conditions above?

2.12 Now think about the meaning of $\psi(x)$. If it is to represent a probability density function, what mathematical condition must be satisfied by $\int |\psi(x)|^2 \, dx$?

   Look again at your answer to question 2.11 together with fig. 2.10. Try to explain why the improper wavefunctions cannot give rise to probability density functions. What is the connection with the three conditions above?

2.8   Normalisation and orthogonality
   We have seen that Born said that if ψ was a solution of the Schrödinger equation then $|\psi|^2$ is a probability density function and

$$\int |\psi|^2 \, dv = 1.$$

Wavefunctions which obey this rule are normalised (or, strictly normalised to unity). We have used the symbol dv to represent a small volume of space because we must assume that in general ψ can spread in three dimensions.

   In section 2.3 the wavefunction for a particle in a one-dimensional box was written $\psi_n = A \sin(n\pi x/l)$. Later we replaced the amplitude, A, by $\sqrt{(2/l)}$. The reason for this was to ensure that the ψ_n were normalised. We have

$$\int_0^l |\psi_n|^2 \, dx = \int_0^l A^2 \sin^2(n\pi x/l) \, dx$$

$$= A^2 \int_0^l \sin^2(n\pi x/l) \, dx,$$

and we can show that the value of this integral is $l/2$ (M.2.8). Thus,

$$A^2(l/2) = 1$$

which gives

$$A = \pm \sqrt{(2/l)}.$$

Because the probability density depends on the

Fig. 2.10. The sketches show a mixture of possible and impossible wavefunctions.

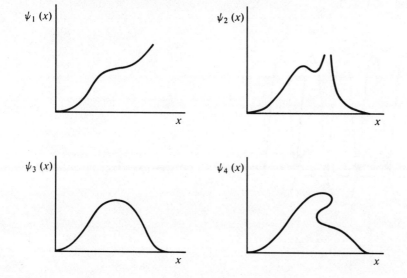

square of ψ_n, which will always be positive, it does not matter if the wavefunction itself is negative or positive in sign. However, for convenience we chose the positive square root.

There is a further property of the wavefunctions that we should examine. This is that if ψ_1 and ψ_2 are different wavefunctions

$$\int \psi_1 \psi_2 \, d\upsilon = 0.$$

This says that the values of $\psi_1 \psi_2$ at all points in space will add together so that the positive values are exactly cancelled out by the negative values.

For example, if we look at graphs of ψ_1 and ψ_2 for the particle in a one-dimensional box we can see that $\psi_1 \psi_2$ is symmetrical about $x = l/2$ (fig. 2.11). To the left of the midpoint, $\psi_1 \psi_2$ is always positive, to the right always negative. The symmetry of the figure shows that the two regions will cancel out if added together.

An integral such as $\int \psi_1 \psi_2 d\upsilon$ is known as an overlap integral. A glance at fig. 2.11 should give an idea of why this is so. If the integral is indeed zero then we can either say that ψ_1 and ψ_2 are orthogonal, or that there is no overlap between them. The former choice emphasises their mathematical properties, the latter a more pictorial approach.

Fig. 2.11. The diagram shows how the wavefunctions ψ_1, ψ_2 and their product vary across the length of a one-dimensional box. The positive area under $\psi_1 \psi_2$ is cancelled out by an equal negative area.

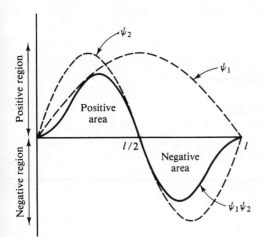

Not all overlap integrals are zero because not all wavefunctions are orthogonal. This is especially important in theories of chemical bonding for, as we shall, see, if there is a good overlap between the wavefunctions belonging to the atoms in a molecule then there is a good chance of strong chemical bonds being formed.

2.9 Summary

We have seen that in order to determine the energy levels of a system we have to write down the Schrödinger equation and solve it. The solutions are the wavefunctions of the system. Owing to the conditions placed on the wavefunctions, quantum numbers appear quite naturally in the solutions for the energy.

In our investigation of the particle in a box problem, zero point energy made its appearance; a result not to be found in classical theory. It turned out that there may be more than one way of achieving a particular value of the energy, in which case the solutions of the Schrödinger equation are degenerate. We noted that we should expect to find degenerate solutions for the energy of an electron in the hydrogen atom because degeneracy reflects the symmetry of the potential well (or box). We also found that the principle of superposition allowed us to create new solutions of the Schrödinger equation by taking linear combinations of the original solutions.

Modern quantum theory relies heavily on Born's interpretation of the wavefunction. This was that $|\psi|^2$ behaves as a probability density function. We can no longer be completely certain of the whereabouts of the electron. The best we can do is to say in which region of space it is most, or least, likely to be found. Modern quantum theory is therefore a statistical theory dealing with uncertainties, unlike classical theory which deals in certainties.

The fact that we must have $\int |\psi|^2 \, d\upsilon = 1$ means that wavefunctions should be normalised. For two orthogonal wavefunctions, $\int \psi_1 \psi_2 \, d\upsilon = 0$ and there is no overlap between them. These two properties are very important for they lie behind our understanding of chemical bonding. This is the subject of the next three chapters.

M.2.1

We need to know that

$$\frac{d(\sin kx)}{dx} = k \cos kx$$

and

$$\frac{d(\cos kx)}{dx} = -k \sin kx.$$

Thus,

$$\frac{d\psi}{dx} = \frac{2\pi}{\lambda} A \cos\left(\frac{2\pi x}{\lambda}\right)$$

$$\frac{d^2\psi}{dx^2} = -\frac{4\pi^2}{\lambda^2} A \sin\left(\frac{2\pi x}{\lambda}\right)$$

$$= -\frac{4\pi^2}{\lambda^2}\psi.$$

Using our previous result for λ,

$$\frac{d^2\psi}{dx^2} = -\frac{4\pi^2 \cdot 2m_e E}{h^2}\psi$$

which gives the result.

Here we can pause a moment to reflect on what happens when the electron can move in the y-direction as well as the x-direction. We would then expect ψ to be given by

$$\psi = A \sin\left(\frac{2\pi x}{\lambda}\right) \sin\left(\frac{2\pi y}{\lambda}\right)$$

and the Schrödinger equation to be

$$-\frac{\hbar^2}{2m_e}\left(\frac{d^2\psi}{dx^2} + \frac{d^2\psi}{dy^2}\right) = E\psi.$$

However, this would be incorrect. Rather we should put

$$-\frac{\hbar^2}{2m_e}\left(\frac{\partial^2\psi}{\partial x^2} + \frac{\partial^2\psi}{\partial y^2}\right) = E\psi.$$

An apparently slight change in notation has occurred, with d^2/dx^2, d^2/dy^2 being replaced by the *partial differentials* $\partial^2/\partial x^2$, $\partial^2/\partial y^2$. The differential $\partial/\partial x$ acting on a function means 'differentiate with respect to x while keeping all other variables constant'. Analogous statements hold for $\partial/\partial y$ and $\partial/\partial z$. For example, $\partial/\partial x$ acting on $x^3 y^2$ gives $3x^2 y^2$. Similarly $\partial^2/\partial y^2$ acting on $x^3 y^2$ gives $2x^3$. Although the notation changing from 'ordinary' differentials, such as d^2/dx^2, to partial differentials, such as $\partial^2/\partial x^2$, appears slight, the change in meaning is essential. We shall find partial differentials occur frequently from now on. This is because we shall spend most of the time dealing with electrons, atoms and molecules which exist in three-dimensional space. Therefore the Schrödinger equation and its solutions will vary with (at least) the three spatial coordinates x, y and z.

M.2.2

For a two-dimensional box we have

$$-\frac{\hbar^2}{2m}\frac{\partial^2\psi}{\partial x^2} - \frac{\hbar^2}{2m}\frac{\partial^2\psi}{\partial y^2} = E\psi.$$

But

$$\frac{\partial^2 X(x)}{\partial x^2} = -\frac{n_x^2 \pi^2}{l^2} X(x)$$

and

$$\partial^2 X(x)/\partial y^2 = 0$$

because $X(x)$ is independent of y. Thus

$$\frac{h^2 n_x^2}{8ml^2} X(x) - 0 = E X(x)$$

so

$$E = \frac{h^2 n_x^2}{8ml^2}$$

which shows that $X(x)$ is still a solution of the Schrödinger equation.

M.2.3

To show that $\psi(x, y)$ is a solution of the Schrödinger equation, we shall first revise the rule for the differentiation of a product i.e.

$$\frac{\partial(uv)}{\partial x} = v\frac{\partial u}{\partial x} + u\frac{\partial v}{\partial x}.$$

Now

$$\frac{\partial \psi(x, y)}{\partial x} = \frac{\partial [X(x)\,Y(y)]}{\partial x}$$

$$= Y(y)\frac{\partial X(x)}{\partial x} + X(x)\frac{\partial Y(y)}{\partial x}$$

but because $Y(y)$ is independent of x, we have $\partial Y(y)/\partial x = 0$. Thus,

$$\frac{\partial \psi(x, y)}{\partial x} = Y(y)\frac{\partial X(x)}{\partial x}$$

and

$$\frac{\partial^2 \psi(x, y)}{\partial x^2} = Y(y)\frac{\partial^2 X(x)}{\partial x^2}$$

$$= -\frac{\pi^2 n_x^2}{l^2}\psi(x, y).$$

(See M.2.2.) Similarly,

$$\frac{\partial^2 \psi(x, y)}{\partial y^2} = -\frac{\pi^2 n_y^2}{l^2}\psi(x, y).$$

Then substituting into the Schrödinger equation gives

$$-\frac{\hbar^2}{2m}\left[-\frac{\pi^2 n_x^2}{l^2}\psi(x, y) - \frac{\pi^2 n_y^2}{l^2}\psi(x, y)\right]$$

$$= E\psi(x, y)$$

which cancels down to produce

$$E_{n_x, n_y} = (n_x^2 + n_y^2)\frac{h^2}{8ml^2}.$$

M.2.4

As with most tasks, there is a long and a short way of showing that ψ_+ has the same energy as $\psi_1(x, y)$ or $\psi_2(x, y)$. The long way is to substitute the precise forms of the wavefunctions into the Schrödinger equation and do the differentiation. This is straightforward but tedious. We shall adopt the short method here.

First we shall write the Schrödinger

equation in the form

$$-\frac{\hbar^2}{2ml^2}\left(\frac{\partial^2}{\partial x^2} + \frac{\partial^2}{\partial y^2}\right)\psi(x, y) = E\psi(x, y)$$

and we shall substitute the symbol $\hat{H}$ for the terms involving the differentials. The 'hat' (^) on H implies that it hides $\partial^2/\partial x^2$ and $\partial^2/\partial y^2$ within it. $\hat{H}$ is an example of an operator, but more of this in chapter 6. We now have

$$\hat{H}\psi(x, y) = E\psi(x, y).$$

Because the energy of $\psi_1(x, y)$ is E_1 and of $\psi_2(x, y)$ is E_2,

$$\hat{H}\psi_1(x, y) = E_1\psi_1(x, y)$$
$$\hat{H}\psi_2(x, y) = E_2\psi_2(x, y)$$

so

$$\hat{H}[\psi_1(x, y) + \psi_2(x, y)]$$
$$= E_1\psi_1(x, y) + E_2\psi_2(x, y).$$

However, we have assumed that $\psi_1(x, y)$ and $\psi_2(x, y)$ are degenerate. This means that $E_1 = E_2$, which allows us to put

$$\hat{H}[\psi_1(x, y) + \psi_2(x, y)]$$
$$= E_1[\psi_1(x, y) + \psi_2(x, y)]$$

i.e.

$$\hat{H}\psi_+ = E_1\psi_+.$$

This completes the task because we have shown that the energy of ψ_+ is just the same as $\psi_1(x, y)$ or $\psi_2(x, y)$.

M.2.5

If a fair die is thrown, the possible outcomes are the integers 1, 2, 3, 4, 5, or 6. These results form a discrete distribution; clearly the results cannot merge into one another. There is an equal probability, $\frac{1}{6}$, of each result occurring:

$$P_1 = P_2 = P_3 = P_4 = P_5 = P_6 = \tfrac{1}{6}$$

so

$$\sum P_i = \tfrac{1}{6} + \tfrac{1}{6} + \tfrac{1}{6} + \tfrac{1}{6} + \tfrac{1}{6} + \tfrac{1}{6}$$

$$= 1.$$

Now as required. The mean is

$$\mu = \sum x_i P_i$$
$$= 1 \times \tfrac{1}{6} + 2 \times \tfrac{1}{6} + 3 \times \tfrac{1}{6} + 4 \times \tfrac{1}{6}$$
$$\quad + 5 \times \tfrac{1}{6} + 6 \times \tfrac{1}{6}$$
$$= 3\tfrac{1}{2}.$$

This illustrates the general point that the mean of a series of results is not necessarily one of the actual outcomes.

M.2.6

To see how we arrive at the form of the classical probability density for the particle in a one-dimensional box, first let us imagine the box cut into halves. We would expect there to be a probability of 0.5 of finding the particle in each half. Similarly, if the box were cut into quarters, the corresponding probability would be 0.25. This suggests that

$$\begin{pmatrix} \text{Probability of finding} \\ \text{the particle} \end{pmatrix} \propto \begin{pmatrix} \text{Length of box} \\ \text{cut off} \end{pmatrix}$$

If we split the length into infinitesimally small portions, dx

$$\begin{pmatrix} \text{Probability of finding} \\ \text{the particle} \end{pmatrix} \propto \; dx$$
$$= \; k dx$$

where k is a constant.

We know that the total probability of finding the particle in the box must be one. The total probability is found by summing all the contributions $k \, dx$ i.e.

$$\int_0^l k \, dx = 1$$
$$k[x]_0^l = 1$$

so,

$$k = 1/l$$

and we have shown that $\int_0^l (1/l) \, dx = 1$.

Comparing this with $\int P(x) \, dx = 1$ shows that in this case $P(x)$, the probability density, is given by the constant $1/l$.

M.2.7

We have to show that

$$\frac{2}{l} \int_0^l \sin^2 \left(\frac{n\pi x}{l} \right) dx = 1.$$

This can be done by first using the identity

$$\sin^2 kx = \tfrac{1}{2}(1 - \cos 2kx)$$

so, with $k = n\pi/l$,

$$\frac{2}{l} \int_0^l \sin^2 \left(\frac{n\pi x}{l} \right) dx$$
$$= \frac{1}{l} \int_0^l (1 - \cos 2kx) \, dx$$
$$= \frac{1}{l} \int_0^l dx - \frac{1}{l} \int_0^l \cos 2kx \, dx$$
$$= \frac{1}{l} [x]_0^l - \frac{1}{l} \left[\frac{1}{2k} \sin 2kx \right]_0^l$$
$$= 1 - \frac{1}{2n\pi} \left[\sin \left(\frac{2n\pi x}{l} \right) \right]_0^l.$$

When $x = l$, $\sin(2n\pi x/l) = \sin 2n\pi$, which is always zero. Of course, $\sin(2n\pi x/l) = 0$ when $x = 0$. Thus we obtain the required result.

M.2.8

The method of showing that

$$\int_l^0 \sin^2 \left(\frac{n\pi x}{l} \right) dx = \frac{l}{2}$$

follows from M.2.7 if we remove the factor $\sqrt{(2/l)}$ from $\sqrt{(2/l)} \sin(n\pi x/l)$ and replace it by A. Then

$$\int_0^l |\psi_n|^2 \, dx = A^2 \int_0^l \tfrac{1}{2}[1 - \cos(2n\pi x/l)] \, dx$$
$$= A^2 [\tfrac{1}{2} x]_0^l$$

because the integral over $\cos(2n\pi x/l)$ is zero. Thus if we want ψ_n to be normalised,

$$1 = A^2 \frac{l}{2}$$

as we have claimed.

3

Electrons in atoms

3.1 Introduction

In this chapter we shall examine the solutions of the Schrödinger equation for the hydrogen atom. In order to make the Schrödinger equation as simple as possible we shall ignore the kinetic energy associated with the movement of the atom as a whole, i.e. its translational energy. Also, because the proton at the nucleus is some 2000 times more massive than the electron, we shall assume that the nucleus remains stationary while the electron orbits about it. This leaves us with just two terms:

(i) the kinetic energy of the electron,

$$-\frac{\hbar^2}{2m_e}\left(\frac{\partial^2}{\partial x^2}+\frac{\partial^2}{\partial y^2}+\frac{\partial^2}{\partial z^2}\right)\psi(x,y,z),$$

(ii) the potential energy, $V(x,y,z)\psi(x,y,z)$ where

$$V(x,y,z)=-\frac{e^2}{4\pi\varepsilon_0}\frac{1}{\sqrt{(x^2+y^2+z^2)}}.$$

This gives us the overall Schrödinger equation

$$-\frac{\hbar^2}{2m_e}\left(\frac{\partial^2}{\partial x^2}+\frac{\partial^2}{\partial y^2}+\frac{\partial^2}{\partial z^2}\right)\psi(x,y,z)$$

$$-\frac{e^2}{4\pi\varepsilon_0}\frac{1}{\sqrt{(x^2+y^2+z^2)}}\psi(x,y,z)$$

$$=E\psi(x,y,z).$$

Having written down the equation it is quite another thing to solve it. Indeed, in this form it cannot be solved exactly. The reason for this is that, unlike the case of the particle in a two- or three-dimensional box, it is not possible to separate the potential energy term into three separate parts, one depending only on x, one only on y, and one only on z. However, all is not lost because it does become possible to solve the equation if we change from the Cartesian coordinate system to the polar coordinate system (fig. 3.1).

In this system we describe the position of the electron in terms of its distance, r, from the nucleus and two angles θ and ϕ. When we look at the Schrödinger equation in spherical polar coordinates it seems we have made matters worse rather than better:

$$-\frac{\hbar^2}{2m_e}\left\{\frac{1}{r^2}\frac{\partial}{\partial r}\left[r^2\frac{\partial\psi(r,\theta,\phi)}{\partial r}\right]\right.$$

$$+\frac{1}{r^2\sin\theta}\frac{\partial}{\partial\theta}\left[\sin\theta\frac{\partial\psi(r,\theta,\phi)}{\partial\theta}\right]$$

$$+\left.\frac{1}{r^2\sin\theta}\frac{\partial^2\psi(r,\theta,\phi)}{\partial\phi^2}\right\}$$

$$+V(r)\psi(r,\theta,\phi)=E\psi(r,\theta,\phi)$$

where $V(r)=-e^2/4\pi\varepsilon_0 r$.

Appearances are deceptive though, for the equation separates nicely into three parts (M.3.1) one depending on r, one on θ, and one on ϕ. Similarly we can isolate three separate contributions to the total wavefunction:

$$\psi(r,\theta,\phi)=R(r)\Theta(\theta)\Phi(\phi).$$

$R(r)$ is the radial part of the wavefunction while $\Theta(\theta)$ and $\Phi(\phi)$ define the angular parts. In addition we find in the solutions the arrival of three quantum numbers. The radial part involves just one of them called the principal quantum number, n. The angular parts involve the other two. These are the azimuthal quantum number, l, and the magnetic quantum number, m_l.

We could show the wavefunctions with their quantum numbers:

$$\psi_{n,l,m_l}(r,\theta,\phi)=R_{n,l}(r)\Theta_{l,m_l}(\theta)\Phi_{m_l}(\phi),$$

but to avoid all the subscripts we shall usually omit them.

In the next section we shall take a closer look at the wavefunctions and the three quantum numbers.

Fig. 3.1. Cartesian and spherical polar co-ordinates: (a) Cartesian coordinates; (b) spherical polar coordinates. The relation between the two systems is

$$x = r \sin \theta \cos \phi$$
$$y = r \sin \theta \sin \phi$$
$$z = r \cos \theta.$$

The small volume element of space $dx dy dz$ in Cartesian coordinates becomes $r^2 \sin \theta dr d\theta d\phi$ in spherical polar coordinates.

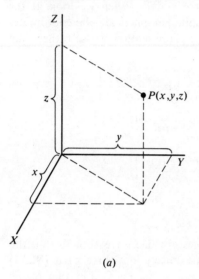

(a)

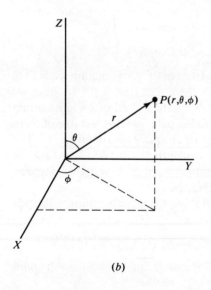

(b)

3.2 Orbitals and quantum numbers

Each wavefunction, $\psi(r, \theta, \phi)$, which is a solution of the Schrödinger equation is known as an atomic orbital, or just orbital for short. Each orbital has a particular set of values of the three quantum numbers n, l, m_l associated with it (M.3.2). The principal quantum number determines the energy of the orbital:

$$E_n = -\frac{m_e e^4}{8\varepsilon_0^2 h^2} \frac{1}{n^2}.$$

We can see that the solution that Schrödinger found for E exactly reproduced Bohr's result (section 1.5).

The azimuthal quantum number gives us information about the shape of the orbital and, of more importance, the orbital angular momentum of the electron. In fact Schrödinger's solution showed that the values of the total orbital angular momentum, L, were quantised according to the rule

$$L = \sqrt{[l(l + 1)]}\hbar.$$

Notice the difference between this and Bohr's result which predicted that L should be given by an equation of the form $L = l\hbar$.

The magnetic quantum number, m_l, tells us the number of orbitals with a given value of l and gives information about what happens to their energies when the symmetry of the atom is disturbed, for example by a magnetic or electric field placed across the atom. We should not be surprised to learn that because of the high symmetry of the potential energy well, many of the orbitals are degenerate. In fact, the degree of degeneracy is given by the square of the principal quantum number. All three quantum numbers are involved in the way the orbitals are named. For convenience much of the information we shall need is collected together in tables 3.1, 3.2, and 3.3.

The way that the orbitals spread in space is of extreme interest because we use this information in theories of chemical bonding. It is not at all easy to show the way the orbitals behave in three dimensions when we are confined to drawing structures in two dimensions. As a result of this difficulty, several tactics can be employed to help us visualise the orbitals. In particular we shall separate the wavefunctions into their orbital and angular parts and look at each individually. To illustrate the method

Table 3.1 *The quantum numbers n, l, and m_l*

Symbol	Name	Information
n	Principal	Governs the orbital energy, $$E_n = -\frac{m_e e^4}{8\varepsilon_0^2 h^2}\frac{Z^2}{n^2},$$ and degeneracy, n^2. Z is the atomic number.
l	Azimuthal	Governs the orbital shape and total angular momentum $$L = \sqrt{[l(l+1)]}\hbar.$$
m_l	Magnetic	Governs the number of orbitals for each value of l, and their behaviour in magnetic or electric fields.

Note: Because m_l and l are related, the magnetic quantum number also gives information about angular momentum. We shall turn to this in chapter 5.

Table 3.2. *Orbitals and values of n, l, m_l*

Orbital	n	l	m_l	Degeneracy (n^2)
1s	1	0	0	1
2s	2	0	0	
$2p_x$	2	1	± 1	
$2p_z$	2	1	0	4
$2p_y$	2	1	± 1	
3s	3	0	0	
$3p_x$	3	1	± 1	
$3p_z$	3	1	0	
$3p_y$	3	1	± 1	
$3d_{x^2-y^2}$	3	2	± 2	9
$3d_{xz}$	3	2	± 1	
$3d_{z^2}$	3	2	0	
$3d_{yz}$	3	2	± 1	
$3d_{xy}$	3	2	± 2	

For $n = 4$ we find 4s, 4p, 4d and 4f.
For $n = 5$ we find 5s, 5p, 5d, 5f and 5g, and so on, but we shall rarely encounter orbitals with $n \geq 4$.
The relationships between the integers n, l, and m_l are best summarised by

$$n > 0; \quad 0 \leq l \leq n - 1; \quad -l \leq m_l \leq +l.$$

Table 3.3. *The wavefunctions for hydrogen-like atoms*

Orbital name	Wavefunction, $\psi(r, \theta, \phi)$
1s	$\frac{1}{\sqrt{\pi}}\left(\frac{Z}{a_0}\right)^{3/2} e^{-\rho}$
2s	$\frac{1}{4\sqrt{(2\pi)}}\left(\frac{Z}{a_0}\right)^{3/2}(2-\rho)e^{-\rho/2}$
$2p_z$	$\frac{1}{4\sqrt{(2\pi)}}\left(\frac{Z}{a_0}\right)^{3/2}\rho e^{-\rho/2}\cos\theta$
$2p_x$	$\frac{1}{4\sqrt{(2\pi)}}\left(\frac{Z}{a_0}\right)^{3/2}\rho e^{-\rho/2}\sin\theta\cos\phi$
$2p_y$	$\frac{1}{4\sqrt{(2\pi)}}\left(\frac{Z}{a_0}\right)^{3/2}\rho e^{-\rho/2}\sin\theta\sin\phi$
3s	$\frac{1}{81\sqrt{(3\pi)}}\left(\frac{Z}{a_0}\right)^{3/2}(27-18\rho+2\rho^2)e^{-\rho/3}$
$3p_z$	$\frac{2}{81\sqrt{(2\pi)}}\left(\frac{Z}{a_0}\right)^{3/2}(6-\rho)\rho e^{-\rho/3}\cos\theta$
$3p_x$	$\frac{2}{81\sqrt{(2\pi)}}\left(\frac{Z}{a_0}\right)^{3/2}(6-\rho)\rho e^{-\rho/3}\sin\theta\cos\phi$
$3p_y$	$\frac{2}{81\sqrt{(2\pi)}}\left(\frac{Z}{a_0}\right)^{3/2}(6-\rho)\rho e^{-\rho/3}\sin\theta\sin\phi$
$3d_{z^2}$	$\frac{1}{81\sqrt{(6\pi)}}\left(\frac{Z}{a_0}\right)^{3/2}\rho^2 e^{-\rho/3}(3\cos^2\theta-1)$
$3d_{xz}$	$\frac{2}{81\sqrt{(2\pi)}}\left(\frac{Z}{a_0}\right)^{3/2}\rho^2 e^{-\rho/3}\sin\theta\cos\theta\cos\phi$
$3d_{yz}$	$\frac{2}{81\sqrt{(2\pi)}}\left(\frac{Z}{a_0}\right)^{3/2}\rho^2 e^{-\rho/3}\sin\theta\cos\theta\sin\phi$
$3d_{xy}$	$\frac{1}{81\sqrt{(2\pi)}}\left(\frac{Z}{a_0}\right)^{3/2}\rho^2 e^{-\rho/3}\sin^2\theta\sin 2\phi$
$3d_{x^2-y^2}$	$\frac{1}{81\sqrt{(2\pi)}}\left(\frac{Z}{a_0}\right)^{3/2}\rho^2 e^{-\rho/3}\sin^2\theta\cos 2\phi$

The quantity ρ is given by

$$\rho = Zr/a_0$$

where Z is the atomic number.

we shall choose the 1s, 2s, and $2p_z$ orbitals, whose wavefunctions we shall write as ψ_{1s}, ψ_{2s}, and ψ_{2p_z} (table 3.4). The radial parts of the wavefunctions are shown in fig. 3.2.

The 1s orbital has the lowest energy and has no nodes while the 2s (and 2p) orbital of higher energy has a single node. This reflects our discussion in section 2.2. As the energy increases, the curvature of the wavefunction increases, but the area under the wavefunction must not become infinite (otherwise the total probability density could never be finite). Thus the wavefunction must always 'turn over'

Table 3.4. *Details of 1s, 2s, and 2p$_z$ wavefunctions of hydrogen*

orbital	$R(r)$	$\Theta(\theta)$	$\Phi(\phi)$	$\psi(r,\theta,\phi)$
1s	$2\left(\dfrac{1}{a_0}\right)^{3/2}e^{-r/a_0}$	$\dfrac{\sqrt{2}}{2}$	$\dfrac{1}{\sqrt{(2\pi)}}$	$\dfrac{1}{\sqrt{\pi}}\left(\dfrac{1}{a_0}\right)^{3/2}e^{-r/a_0}$
2s	$\dfrac{1}{2\sqrt{2}}\left(\dfrac{1}{a_0}\right)^{3/2}\left(2-\dfrac{r}{a_0}\right)e^{-r/2a_0}$	$\dfrac{\sqrt{2}}{2}$	$\dfrac{1}{\sqrt{(2\pi)}}$	$\dfrac{1}{4\sqrt{(2\pi)}}\left(\dfrac{1}{a_0}\right)^{3/2}\left(2-\dfrac{r}{a_0}\right)e^{-r/2a_0}$
2p$_z$	$\dfrac{1}{2\sqrt{6}}\left(\dfrac{1}{a_0}\right)^{3/2}\dfrac{r}{a_0}e^{-r/2a_0}$	$\dfrac{\sqrt{6}}{2}\cos\theta$	$\dfrac{1}{\sqrt{(2\pi)}}$	$\dfrac{1}{4\sqrt{(2\pi)}}\left(\dfrac{1}{a_0}\right)^{3/2}\dfrac{r}{a_0}e^{-r/2a_0}\cos\theta$

a_0 is the first Bohr radius: $a_0 = \varepsilon_0 h^2/\pi m_e e^2$. Atoms, or ions, with atomic number Z but with just one electron have wavefunctions of exactly the same symmetry but with $1/a_0$ replaced by Z/a_0.

Each orbital has energy

$$E_n = -\frac{m_e e^4}{8\varepsilon_0^2 h^2}\frac{1}{n^2},$$

or, more generally

$$E_n = -\frac{m_e e^4}{8\varepsilon_0^2 h^2}\frac{Z^2}{n^2}.$$

towards the r-axis, bringing a tendency to cross it and produce a node. We should remember that because of the three-dimensional nature of the orbitals, the nodes are likely to form nodal surfaces. For example, the 2s orbital has a spherical nodal surface at a radius of $2a_0$. In the sphere of radius less than $2a_0$ the wavefunction is positive in sign; beyond $2a_0$ it is negative. With some thought you should be

able to imagine the three-dimensional behaviour of the radial parts of the 1s and 2p$_z$ orbitals.

To understand diagrams of angular parts of the wavefunctions we shall turn to the 2p$_z$ orbital as an example. For this orbital the angular part is proportional to $\cos\theta$. We draw lines out from the origin at angles θ_1, θ_2 as shown in fig. 3.3(a). Then we mark off on each one a length proportional to

Fig. 3.2. Diagrams for the radial wavefunctions, $R(r)$, for hydrogen 1s, 2s, and 2p orbitals. Note that the diagrams are not to the same scale. The units of $R(r)$ are $(1/a_0)^{3/2}$.

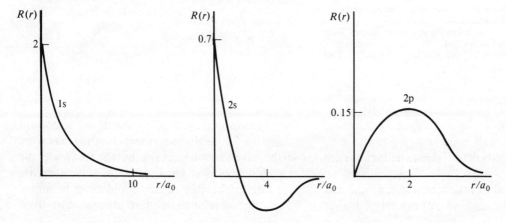

$\cos\theta_1$, $\cos\theta_2$. When this is done for θ between 0 and $\pi/2$ and between $3\pi/2$ and 2π, $\cos\theta$ is always positive and we obtain a circle (fig. 3.3(b)). The fact that $\cos\theta$ is positive is shown by writing a plus sign in the circle.

Fig. 3.3. Representing the angular part of a $2p_z$ wavefunction. (a) The distance of each point from the origin is proportional to $\cos\theta$. (b) The positive sign shows that $\cos\theta$ is positive for $0 \leqslant \theta < \pi/2$ and $3\pi/2 \leqslant \theta < 2\pi$. (c) For $\pi/2 \leqslant \theta < 3\pi/2$, $\cos\theta$ is negative as shown by the sign in the lower lobe.

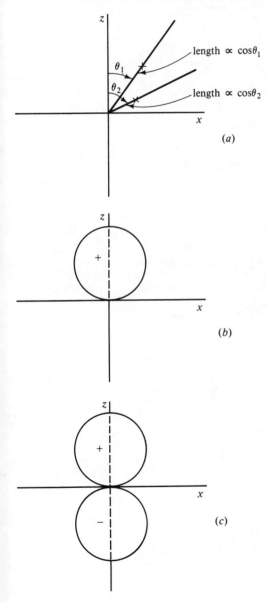

When θ is between $\pi/2$ and $3\pi/2$, $\cos\theta$ is negative, but the sign is ignored in marking off the lengths as before, although it is placed in the circle. Thus we obtain the final diagram in fig. 3.3(c). The diagrams for 1s and 2s orbitals are identical as their angular parts are both constants independent of θ or ϕ. The diagrams for $2p_x$ and $2p_y$ orbitals are of the same symmetry as that of the $2p_z$ except that they are oriented about the x- and y-axes respectively. Indeed any set of p orbitals have the same angular symmetry as the 2p set and all s orbitals have the same angular symmetry as the 1s (fig. 3.4).

Finally we should bring the radial and angular parts of the wavefunctions together. Again taking the $2p_z$ orbital as an example, if we choose a value for r and then let θ vary we can tabulate a series of values for the wavefunction. This procedure can be repeated for further values of r until we have sufficient information to draw a set of contours joining points at which the wavefunction has a given value. In fig. 3.5 are shown a selection of contours for cross-sections of the 1s, 2s and $2p_z$ orbitals. Only a few contours are drawn. As the wavefunction spreads out to infinity, by rights the contours should spread in similar fashion. Of course, a halt has to be called once the values of the wavefunctions become very small.

Questions

3.1 If you have access to a computer, write a program to calculate values for the 1s wavefunction of H, He^+, Li^{2+}, Be^{3+}, B^{4+}. Even better, adapt your program to plot graphs of each wavefunction. How pronounced is the effect of increasing nuclear charge on the spread of the orbitals?

To do this question you need to know that Schrödinger's method can be applied to any element, of nuclear charge Ze, provided it has only one electron. Schrödinger obtained the same result that you should have found in answer to question 1.8:

$$E_n = -\frac{m_e e^4}{8\varepsilon_0^2 h^2}\frac{Z^2}{n^2}.$$

Also, note that we can not easily apply the method of this problem to orbitals other than s orbitals. This is because p, d, f, ... orbitals are

Fig. 3.4. Diagrams of angular wavefunctions for s and p orbitals: (a) s orbitals all have the same spherical symmetry; (b) for p orbitals the angular parts form two circular lobes touching at the origin.

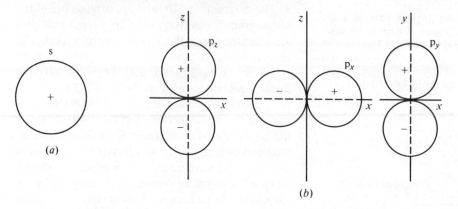

Fig. 3.5. Contour diagrams of $\psi(r, \theta, \phi)$ for 1s, 2s and 2p orbitals. (Not to scale.) Note that there is a nodal surface for the 2s orbital where the wavefunction changes sign on passing through zero. As usual, one lobe of a 2p orbital has positive values for its wavefunction and the other lobe has negative values.

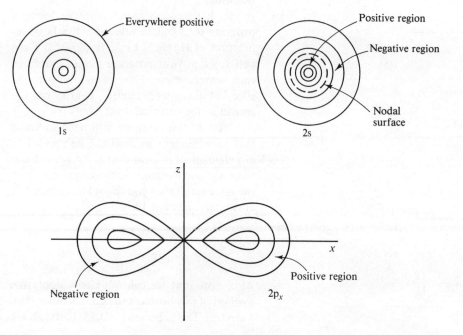

not spherically symmetric like s orbitals. We shall meet the importance of this shortly.

3.3 Probability density diagrams

In the previous chapter we saw that the wavefunction alone could give us no useful inform-

Matthews *Quantum chemistry of atoms and molecules*

ISBN 0 521 24854 x hard covers
ISBN 0 521 27025 1 paperback

Erratum (1986 printing)

p. 34, Fig. 3.5, the correct form for the contour diagram of $\psi(r, \theta, \phi)$ for the 2p orbital is:

Negative region

Positive region

$2p_x$

δr

the square of
a probability
to check that
rmalised and
be the case in
hat the radial
alised accord-

(A)

$|^2$ as a radial
he spherically
ay of writing

(B)

s as it can help
radial proba-

hell of radius, r,
the spheres of
$r + \delta r)^3$ respec-
hickness δr is

$\delta r)^2 + (\delta r)^3$].

$\rightarrow$ dr and we can
hus the volume

Earlier we saw that for a particle in a one-dimensional box, the probability of finding the particle in a length dx was just $|\psi(x)|^2$dx. We might then think that the corresponding quantity for the hydrogen atom would be $|\psi(r, \theta, \phi)|^2dr$, but this ignores the three-dimensional nature of the atom. We need the probability of finding the electron in a given volume. Instead of cutting off a length dr along a single radius, we wish to calculate the probability of finding the electron in a spherical shell of thickness dr (fig. 3.6). This volume is $4\pi r^2$dr. Thus the probability of finding the electron in this shell is $4\pi r^2 |\psi(r, \theta, \phi)|^2dr$. As we have said, equation (B) is only valid for s orbitals and it is better to use $r^2|R(r)|^2$ as the radial probability density function because this works for all orbitals. The factor 4π is absent because this is associated with the angular wavefunctions $\Theta(\theta)$ and $\Phi(\phi)$, but the factor r^2

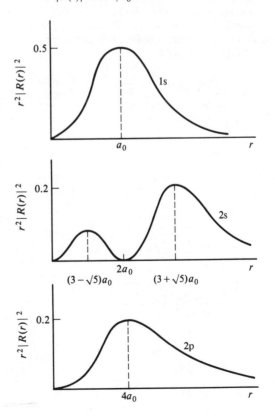

Fig. 3.7. Graphs of radial probability density functions for 1s, 2s, and 2p orbitals. Note: the diagrams are not drawn to the same scale. The units of $r^2|R(r)|^2$ are $1/a_0$.

remains to take account of the spherical shell. We can see that if we write $r^2|R(r)|^2 dr$ differently, i.e. as $|R(r)|^2 r^2 dr$ then $r^2 dr$ represents a small volume of space which is weighted by the factor $|R(r)|^2$.

The graphs of the radial probability density functions for 1s, 2s, and 2p orbitals show some important features (fig. 3.7). The maximum in the function for the 1s orbital comes at exactly the value of the first Bohr radius, $a_0 = 5.292 \times 10^{-11}$ m. Whereas Bohr had predicted that the electron would always be found exactly at a distance a_0 from the nucleus, the Born probability model says that the electron is most probably, but not certainly, to be found at this distance. The most probable distances for the 2s and 2p orbitals come at $(3 \pm \sqrt{5})a_0$ and $4a_0$ respectively (see M.3.3). The larger of the two maxima for the 2s orbital comes beyond that of the 2p orbitals, but the presence of the smaller maximum is most important. It means that there is a higher probability of finding an electron in a 2s orbital nearer the nucleus (nearer than a_0 say)

than an electron in a 2p orbital. This type of behaviour has important repercussions on the way the other electrons in an atom may be shielded from the nucleus by the inner electrons. We shall return to this point in section 3.7.

We should not confuse the most probable distance with the average distance of the electron from the nucleus for these two quantities are quite different. The calculations in M.3.4 show that the average distances from the nucleus for electrons in 1s, 2s, and 2p orbitals are $\frac{3}{2}a_0$, $6a_0$, and $5a_0$ respectively.

The angular probability density functions are proportional to $|\Theta(\theta)\,\Phi(\phi)|^2$. They can be represented by employing the same techniques used for illustrating the angular wavefunction. The main point to notice is that in spite of the fact that probability density functions are always positive, the various lobes of the diagrams are shown containing a positive or negative sign. These signs are the signs of the wavefunctions in that region of

Fig. 3.8. Contour diagrams of probability density functions $|\psi(r, \theta, \phi)|^2$ for 1s, 2s and 2p orbitals. The diagrams are not drawn to scale. Note that the signs are the signs of the wavefunctions in the appropriate region of space.

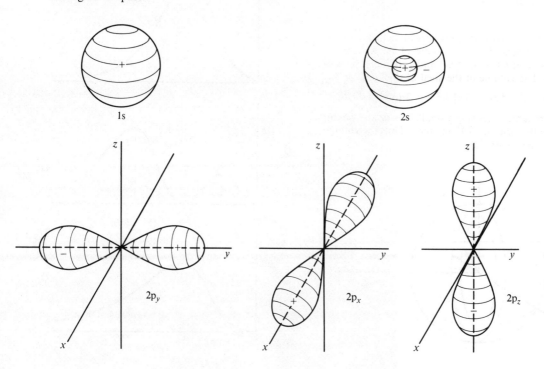

space. This convention will be found useful when we come to discuss the formation of chemical bonds. Contour diagrams of the complete probability density functions are interesting in several respects (fig. 3.8).

In principle the boundary surfaces should extend to infinity but in practice chemists are concerned with where the electron is most likely to be. The boundary surfaces are usually drawn so that the probability of finding an electron within the enclosed region is at least 0.9. Again, we usually include the sign of the wavefunction in the various lobes of the diagrams.

Notice that orbitals other than s orbitals have zero probability density at the nucleus. The probability density function is a maximum at the nucleus for s orbitals, but this does not mean that the probability is a maximum. As we have seen, the probability of finding an electron at a point is zero. On the other hand, it is true that if we take a small volume surrounding the nucleus, we are more likely to find an electron in an s orbital in that volume than an electron in any other type of orbital. This is sometimes paraphrased by saying that electrons in s orbitals can penetrate to the nucleus, while electrons in other orbitals can not.

Questions

3.2 If you know nothing about complex numbers don't attempt this question.

The solutions $\Phi_{m_l}(\phi) = [1/\sqrt{(2\pi)}]e^{im_l\phi}$ involve the complex number $i = \sqrt{(-1)}$. It is shown in standard mathematics textbooks that (the real) cosine and sine functions are formed by taking linear combinations of complex exponentials

$$\cos x = \tfrac{1}{2}(e^{ix} + e^{-ix});$$

$$\sin x = (1/2i)(e^{ix} - e^{-ix}).$$

When $l = 1$, the magnetic quantum number can take the values $m_l = -1, 0, +1$. Use the principle of superposition to obtain the wavefunctions $\Phi(\phi)$ shown in table 3.3 for the 2p orbitals.

3.3 Prove by calculation that the maximum in the radial probability distribution for a 2p orbital does come at $r = 4a_0$ as proposed in M.3.3.

3.4 You may have found that it is quite tedious keeping track of all the constant terms such as h, a_0, m_e, etc. when writing down wavefunctions and performing calculations. There is a standard way round this. It is to use a system of atomic units. In these units, the unit of length becomes the Bohr radius, a_0, and the unit of energy is the magnitude of the energy of the ground state of the hydrogen atom, $m_e e^4/8h^2\varepsilon_0^2$. Similarly, in atomic units $\hbar$ and $4\pi\varepsilon_0$ are set to unity and the electron is given unit mass and charge. The quantity $m_e e^4/8h^2\varepsilon_0^2$ is known as the rydberg unit of energy, E_r. Another, widely used, atomic unit of energy is the hartree, E_h. The relationship between them is very simple: $E_h = 2E_r$.

(i) Convert the Schrödinger equation for the hydrogen atom

$$-\frac{\hbar^2}{2m_e}\nabla^2\psi - \frac{e^2}{4\pi\varepsilon_0 r} = E\psi$$

to atomic units.

(ii) Write down the solution for the energy levels, E_n, in atomic units.

(iii) The wavefunction for a 1s orbital is usually written $(1/\sqrt{\pi})(1/a_0)^{3/2}e^{-r/a_0}$. What is it in atomic units?

3.5 Use the method of M.3.4 to show that r_{av} is $6a_0$ for a 2s orbital and $5a_0$ for a 2p orbital.

3.4 Spin

During the final years of the old quantum theory it became clear that Bohr's theory was not capable of explaining the detailed appearance of the spectra of atoms. In an attempt to resolve some of the problems it was first suggested by A.H. Compton in 1921, and later in more detail by S. Goudsmit and G. Uhlenbeck in 1925 that the angular momentum of an electron was not just due to its orbital motion about the nucleus. It was proposed that each electron had its own intrinsic angular momentum. In classical physics, angular momentum had to be connected with some form of circular motion, so in order to preserve this classical idea it was assumed that the electrons were spinning on their axes. Hence electrons were said to possess spin angular momentum, or just spin for short. The magnitude of the spin was determined by a spin quantum number, s, whose value was restricted to be $\frac{1}{2}$ only.

In modern quantum theory only some aspects of this original theory are retained. We have no evidence for the notion that electrons actually spin. Indeed, the intrinsic angular momentum of an electron cannot be satisfactorily explained by classical physics. However, the quantum number, s, retains its value of $\frac{1}{2}$ and it keeps the name spin quantum number. In common with the formula in section 3.2, for the orbital angular momentum L, the total spin angular momentum, S, is given by

$$S = \sqrt{[s(s+1)]}\hbar.$$

Although the spin quantum number does not appear in the solution of the Schrödinger equation, it does so if relativistic effects are taken into account; but this is not to say that spin is a purely relativistic phenomenon, for it can be derived in other ways.

Just as the orbital angular momentum quantum number, l, gives rise to a further quantum number, m_l, so is s related to another quantum number, m_s. The values of m_s, like those of m_l, become important for explaining the results of experiments in which atoms are placed in magnetic fields. However, m_s can take only two values, $m_s = +\frac{1}{2}$ or $m_s = -\frac{1}{2}$; no other values are allowed. If $m_s = +\frac{1}{2}$, the electron is said to be 'spin-up'; one with $m_s = -\frac{1}{2}$ is 'spin-down'. On diagrams a value of $m_s = +\frac{1}{2}$ may be shown by an arrow ↑ or ⇑. For $m_s = -\frac{1}{2}$, we find ↓ or ⇓.

If we accept that spin is an important quantity, then it is natural to expect the wavefunction of an electron to show a part which governs the spin. To allow this we shall use the symbol α to represent a spin wavefunction for an electron with $s = \frac{1}{2}$, $m_s = +\frac{1}{2}$. Similarly, β will correspond to an electron with $s = \frac{1}{2}$, $m_s = -\frac{1}{2}$. These wavefunctions will be assumed to be normalised and orthogonal according to the rules

$$\int |\alpha|^2 \, \mathrm{d}s = \int |\beta|^2 \, \mathrm{d}s = 1,$$

$$\int \alpha^* \beta \, \mathrm{d}s = \int \beta^* \alpha \, \mathrm{d}s = 0.$$

Here, the integration is imagined to take place over the range of the 'spin variable' s. The star (*) means that the complex conjugate must be taken. See chapter 6 for details of complex numbers.

Finally, we should now remember that an electron in an atom should be associated with five quantum numbers: n, l, m_l, s, m_s. In the next section we shall see how the restrictions that are placed on certain combinations of these quantum numbers lead to an explanation of the electron structures of the elements.

3.5 The Pauli exclusion principle

Wolfgang Pauli published his celebrated principle in 1925. There are two alternative ways of stating the principle. The most powerful version, but the more difficult to understand, we shall leave until section 3.9; the other is more straightforward and will be sufficient for our purposes at present. It says that:

No two electrons in the same atom can have the same set of four quantum numbers n, l, m_l, m_s.

A major outcome of this is that no more than two electrons can exist with the same spatial wavefunction, and that if they do have the same spatial wavefunction then they must have opposite spins. A shorthand way of stating this is that a maximum of two electrons can be 'in' the same orbital and the spins must be 'paired'. Let us take the lithium atom as an example. The ground state wavefunction of lithium is very similar to the 1s orbital of hydrogen, this having values 1, 0, 0 for n, l, m_l respectively. If we assume that the first electron has a spin quantum number $m_s = +\frac{1}{2}$, we have the set 1, 0, 0, $\frac{1}{2}$ for the four quantum numbers. If the second electron goes into the same orbital, it too will

Fig. 3.9. The electron configuration of lithium, $1s^2 2s$.

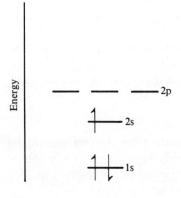

have n, l, m_l as 1, 0, 0. If it is not to have all four quantum numbers the same, it must have $m_s = -\frac{1}{2}$, thus giving the set 1, 0, 0, $-\frac{1}{2}$. When we try to put a third electron into the 1s orbital it would have to possess the same set of four quantum numbers as one of the electrons already there. This is precisely what the Pauli exclusion principle forbids. The third electron would therefore go into the next available orbital, which in the case of lithium would be a 2s orbital. The electron arrangement for lithium could then be shown on a diagram (fig. 3.9).

Before leaving the exclusion principle, we should note that Pauli stated it as a result of his trying to understand the spectra of helium and more complicated atoms. The principle does not actually forbid nature to do anything; but it does tell us what appears to be a fundamental law of nature. It reminds us to keep the book-keeping correct in accounting for the electrons in atoms. Like the Schrödinger equation, its justification lies in its success in rationalising the results of experiments.

Questions

3.6   In order to make sure a wavefunction involving orbital and spin parts is normalised we have to integrate over all space and over the spin coordinate. For the sake of brevity let us write a 2s spatial wavefunction as just 2s. Then if the electron in this orbital has spin α, we must have

$$\int_{\substack{\text{all}\\\text{space}}} \int_{\text{spin}} |2s\alpha|^2 \, ds \, dv = 1$$

or

$$\int_{\substack{\text{all}\\\text{space}}} |2s|^2 dv \int_{\text{spin}} |\alpha|^2 \, ds = 1.$$

Generally this is guaranteed by using correctly normalised functions to begin with.

(i) Explain why the wavefunctions $2s\alpha$ and $2s\beta$ are orthogonal.

(ii) Similarly, why are $2s\beta$ and $2p_z\beta$ orthogonal?

3.7   Why is it not absolutely essential to identify each electron by all five quantum numbers n, l, m_l, s, m_s, when using the Pauli exclusion principle?

3.6   The aufbau method and periodic table

The aufbau method is the method used to build up the electron structures of the elements in the periodic table. The basis of the method we have already met in deciding upon the electron structure of lithium. The key idea is that whenever possible we place electrons in the orbitals of lowest possible energy while still obeying the Pauli exclusion principle. However, there are several things that make the process less straightforward than it might seem.

First, if we return to the case of helium, we said that the ground state wavefunction was similar to that of hydrogen. In fact this is by no means obvious. To determine the wavefunctions for the hydrogen atom it is necessary to solve the Schrödinger equation for the atom. For hydrogen the equation can be solved exactly; but for all other atoms only approximate solutions can be found. To see the reason for this let us look at the helium atom in more detail (fig. 3.10).

Again ignoring the movement of the atom as a whole and assuming the nucleus to be stationary, the terms we have to include in setting up the Schrödinger equation are:

(i) the kinetic energy of the first electron, represented by the operator $-(\hbar^2/2m_e)\nabla_1^2$.

(ii) a similar term for the second electron, $-(\hbar^2/2m_e)\nabla_2^2$.

(iii) the potential energy owing to the attraction between the first electron and the nucleus, $V_1 = -2e^2/4\pi\varepsilon_0 r_1$.

(iv) a similar term for the second electron, $V_2 = -2e^2/4\pi\varepsilon_0 r_2$.

(v) the potential energy owing to the repulsion between the two electrons, $V_{12} = e^2/4\pi\varepsilon_0 r_{12}$.

It is the presence of the final term, V_{12}, which makes the Schrödinger equation impossible to solve exactly. This term is a mathematical statement of the fact that the motions of the two electrons are not

Fig. 3.10. The helium atom consists of a nucleus of charge $+2e$ and two electrons each of charge $-e$. For convenience we have labelled the electrons e_1 and e_2.

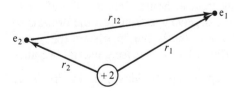

independent of one another. Indeed, as we know, owing to their having identical charges the electrons will tend to avoid finding each other in the same region of space. The motions of the electrons are said to be correlated. If electron correlation is ignored, the Schrödinger equation can be split into two parts:

$$\underbrace{\left(-\frac{\hbar^2}{2m_e}\nabla_1^2 - \frac{2e^2}{4\pi\varepsilon_0 r_1}\right)}_{\substack{\text{kinetic and potential} \\ \text{energies of first electron}}}\psi(r, \theta, \phi)$$

$$+\underbrace{\left(-\frac{\hbar^2}{2m_e}\nabla_2^2 - \frac{2e^2}{4\pi\varepsilon_0 r_2}\right)}_{\substack{\text{kinetic and potential} \\ \text{energies of second electron}}}\psi(r, \theta, \phi)$$

$$= E\psi(r, \theta, \phi).$$

Each part is independent of the other, so each can be solved independently to give exact results for the energies E_1, E_2, and for the wavefunctions $\psi_1(r_1, \theta_1, \phi_1)$, $\psi_2(r_2, \theta_2, \phi_2)$ of the individual electrons. The total energy, E, is just the sum of E_1 and E_2, i.e. $E = E_1 + E_2$. The reason for this success is that the two parts have exactly the same form as the Schrödinger equation for the hydrogen atom but with twice the nuclear charge. This results in the orbitals having exactly the same symmetries as the 1s, 2s, 2p,... orbitals of hydrogen, but with energies four times lower. Analogous results occur for all other atoms if electron correlation is ignored.

Not surprisingly, it can be shown that the effect of electron correlation does make a difference to the way orbitals spread in space, but not such a difference as to destroy the essential similarity of the orbitals to those of hydrogen. For this reason we shall continue to use the terms 1s, 2s, 2p,... as labels for the orbitals of all atoms. Electron correlation does have one major effect, and that is on the energies of the orbitals. In particular the presence of the potential energy term V_{12} corrupts the symmetry of the potential well because now the electrostatic interactions in the atom are no longer directed towards one central point. This has the effect of removing the degeneracies amongst the orbitals so that, for example, the 2s and 2p sets are split apart (fig. 3.11). The order of increasing energy becomes

1s, 2s, 2p, 3s, 3p, 4s, 3d, 4p, 5s, 4d, 5p, 6s,...

We can now develop a better understanding of the aufbau method applied to helium. Clearly, the first of the two electrons will go into a 1s orbital. This electron will have $m_s = +\frac{1}{2}$ or $-\frac{1}{2}$. We do not know which, but we shall assume it has $m_s = +\frac{1}{2}$. The second electron could go into the 1s orbital, but it could go into the 2s. The choice which is made depends on which arrangement minimises the energy. Whenever we place two electrons in the same orbital there will be a contribution to the energy owing to the repulsion between electrons occupying the same region of space. This repulsion tends to increase the energy. If the gap between the 1s and 2s orbitals were very small it could be that the total energy would be lowered by decreasing the repulsions and placing the electrons in different orbitals. In fact the energy gap between the 1s and 2s orbitals is far greater than the repulsion between the two electrons placed in the 1s orbital. Thus the energy is minimised when both electrons enter the 1s orbital. In accordance with the Pauli exclusion principle, the second electron will have $m_s = -\frac{1}{2}$.

By extending this procedure to the other atoms in the periodic table we can deduce their electron structures (table 3.5). You should be able to

Fig. 3.11. In atoms other than hydrogen, the degeneracy of levels with the same principal quantum number is removed. In this diagram the splittings between the energy levels are not to scale.

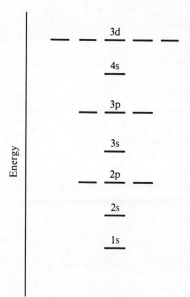

Table 3.5. *The ground state electron structures of the first twenty elements in the periodic table*

Element	1s	2s	$2p_x$	$2p_y$	$2p_z$	3s	$3p_x$	$3p_y$	$3p_z$	4s	Overall
H	↑										$1s$
He	↑↓										$1s^2$
Li	↑↓	↑									$1s^22s$
Be	↑↓	↑↓									$1s^22s^2$
B	↑↓	↑↓	↑								$1s^22s^22p$
C	↑↓	↑↓	↑	↑							$1s^22s^22p^2$
N	↑↓	↑↓	↑	↑	↑						$1s^22s^22p^3$
O	↑↓	↑↓	↑↓	↑	↑						$1s^22s^22p^4$
F	↑↓	↑↓	↑↓	↑↓	↑						$1s^22s^22p^5$
Ne	↑↓	↑↓	↑↓	↑↓	↑↓						$1s^22s^22p^6$
Na	↑↓	↑↓	↑↓	↑↓	↑↓	↑					$1s^22s^22p^63s$
Mg	↑↓	↑↓	↑↓	↑↓	↑↓	↑↓					$1s^22s^22p^63s^2$
Al	↑↓	↑↓	↑↓	↑↓	↑↓	↑↓	↑				$1s^22s^22p^63s^23p$
Si	↑↓	↑↓	↑↓	↑↓	↑↓	↑↓	↑	↑			$1s^22s^22p^63s^23p^2$
P	↑↓	↑↓	↑↓	↑↓	↑↓	↑↓	↑	↑	↑		$1s^22s^22p^63s^23p^3$
S	↑↓	↑↓	↑↓	↑↓	↑↓	↑↓	↑↓	↑	↑		$1s^22s^22p^63s^23p^4$
Cl	↑↓	↑↓	↑↓	↑↓	↑↓	↑↓	↑↓	↑↓	↑		$1s^22s^22p^63s^23p^5$
Ar	↑↓	↑↓	↑↓	↑↓	↑↓	↑↓	↑↓	↑↓	↑↓		$1s^22s^22p^63s^23p^6$
K	↑↓	↑↓	↑↓	↑↓	↑↓	↑↓	↑↓	↑↓	↑↓	↑	$1s^22s^22p^63s^23p^64s$
Ca	↑↓	↑↓	↑↓	↑↓	↑↓	↑↓	↑↓	↑↓	↑↓	↑↓	$1s^22s^22p^63s^23p^64s^2$

explain why, for example, in going from boron to carbon the extra electron enters a second member of the 2p set rather than joining the electron already present in the same orbital. What is more problematic is why the two electrons in the different 2p orbitals of carbon, or nitrogen, keep the same value of m_s. That is, why do the electrons have parallel spins? The answer to this is by no means simple, but it seems to arise from the way the electrons affect each other's degree of attraction for the nucleus. Calculations suggest that it is not primarily due to the influence of one spin on another, which is suggested by phrases such as 'electrons try to keep their spins parallel'.

The electron structures represented in table 3.5 display some points of interest. The noble gases have electron structures corresponding to the filling of a shell or major subshell. By shell we mean a set of orbitals all possessing the same principle quantum number. Often the shells are known by the letters K, L, M, N,... corresponding to $n = 1, 2, 3, 4,...$. The K shell closes with helium, the L shell with neon, and the M shell is partially full at argon. It is clear that the closure of a shell or subshell corresponds to a highly favoured electron structure. When a shell is incomplete there comes the chance for an atom to lose or gain electrons and thus take part in chemical bonding – a matter we shall take up in later chapters.

Questions

3.8 Draw diagrams, like that of fig. 3.9, for the electron structures of the elements hydrogen to fluorine.

3.9 What would a simple filling of orbitals scheme give for the electron configuration of chromium, atomic number 24? How does this compare with the observed configuration $1s^22s^22p^6 3s^23p^64s3d^5$. Try to give a possible explanation.

3.7 Ionisation energies

The ionisation energy of an atom is defined as the energy change accompanying the removal of an electron from the atom in the gaseous state, i.e. it is the energy change for the reaction

$$A(g) \rightarrow A^+(g) + e^-.$$

This energy change is also called the first ionisation energy to distinguish it from the second, third and further ionisation energies, e.g.

$$A^+(g) \rightarrow A^{2+}(g) + e^- \quad \text{(2nd IE)}$$
$$A^{2+}(g) \rightarrow A^{3+}(g) + e^- \quad \text{(3rd IE)}.$$

The measurement of ionisation energies is important because it gives us information about the energies of the orbitals of an atom. Koopmans' theorem states that the ionisation energy for a given electron is equal in magnitude to the energy of the orbital in which it existed. Strictly the theorem is only approximately true, but it is often a very good approximation. Also it has the virtue of applying to electrons in molecules, not just atoms.

A graph of ionisation energy against atomic number for the first ten elements in the periodic table is shown in fig. 3.12. The main trend going across the period lithium to neon is for the ionisation energies to increase. This is not surprising because the energies of the orbitals decrease in proportion to the square of the atomic number (see table 3.4). When going from the end of one period to the start of another, there is a marked drop in ionisation energy. This results from the property of electrons in orbitals nearer to the nucleus to shield the outer electrons from the full influence of the nuclear charge. The extent to which shielding can occur is illustrated in fig. 3.13. The total electron

density for electrons in the K and L shells show their maxima well before the maximum in the 3s electron density. Thus, although a 3s orbital does penetrate the K and L shells to some extent, for the majority of the time the 3s electron would exist beyond those shells and it would not feel the full effects of the nucleus.

In the case of lithium, its $1s^2$ electron structure represents a spherically symmetric distribution of charge about the nucleus. If we imagine trying to add the further 2s electron to the atom, it would be considerably shielded and therefore its energy would be higher than if shielding were absent. Thus, the energy needed to remove the electron would be less than expected. Hence the decrease in ionisation energy from helium to lithium. The influence of shielding also appears in going from beryllium to boron. However, the change from boron to carbon is not so markedly affected by shielding because the two 2p electrons of carbon are in separate members of the 2p set. Owing to the different orientations of p_x, p_y and p_z orbitals the shielding of one p electron by another is largely unimportant. The rise in ionisation for carbon reflects the decrease in energy of the p orbitals owing to the increased nuclear charge. A similar state of affairs accounts for the rise in ionisation energy for nitrogen.

At first sight it might seem that there should be a corresponding rise for oxygen. However, this would be to ignore the effect of having two electrons

Fig. 3.12. A graph of first ionisation energy against atomic number shows apparent anomalies at boron and oxygen. The numbers in brackets give the values of the ionisation energies in kJ mol^{-1}.

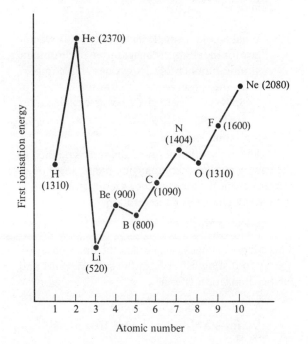

Fig. 3.13. For an atom like sodium the outer 3s electron can penetrate into the region of space occupied by the electrons in the K and L shells. However, for a sodium 3s electron $r_{av} = 1.2a_0$. This is shown on the diagram and is well beyond the bulk of K and L shell electron density.

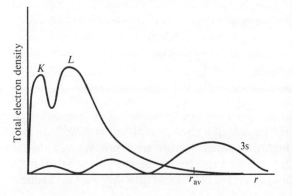

in the same p orbital. As we have already noted, placing two electrons in the same orbital increases the repulsion between them. Thus in spite of the increased nuclear charge we can understand that the drop in ionisation energy reflects the fact that one of the paired electrons is relatively easy to remove owing to its being repelled by its neighbour in the same orbital. The rise from oxygen to neon reflects the balance between increasing nuclear charge, lack of effective shielding by one p orbital of another, and the effect of placing two electrons in the same orbital.

Questions

3.10 If an atom has a low ionisation energy then the electron removed must be in an orbital of high energy, and vice versa. Use Koopmans' theorem to draw a diagram illustrating the orbital energies for the orbitals involved in the first ionisation energy of each of the elements hydrogen to neon.

3.8 Approximate orbitals

If an atom has more than one electron we are forced to develop ways of finding approximate solutions of the Schrödinger equation. We saw in section 3.6 that the simplest approximation is to ignore the influence that one electron has on another. This neglect of electron correlation cannot be justified except on the grounds of expediency. There are a number of different facets to electron correlation. The first, and most obvious, is that owing to their similar charges electrons will avoid being in the same region of space. Breaking this down further, they will tend to avoid being at the same distance along a radius. This is radial correlation. Also they will avoid being at the same angle to the nucleus; indeed, all other things being equal, they will tend to be found on opposite sides of the nucleus. This is angular correlation.

Of greater subtlety is spin correlation, which has its explanation in the sophisticated form of the Pauli exclusion principle. Without going into the detail of this yet, we shall note that, for reasons nothing to do with their charge, electrons with the same spin are unlikely to be found in the same region of space. This phenomenon is one of the many facets of the behaviour of electrons which classical physics finds impossible to explain satisfactorily. Certainly it is not one which can be

Table 3.6. *Zeta-values for Slater orbitals of the first ten elements in the periodic table*

Element	Orbital		
	1s	2s	2p
H	1.0000		
He	1.6875		
Li	2.6906	0.6396	
Be	3.6848	0.9560	
B	4.6795	1.2881	1.2107
C	5.6727	1.6083	1.5679
N	6.6651	1.9237	1.9170
O	7.6579	2.2458	2.2266
F	8.6501	2.5638	2.5500
Ne	9.6421	2.8792	2.8792

explained by thinking of electrons as tiny balls of spinning charge.

A respectable approximate wavefunction for an atom should take account of each of the three types of correlation. However, to do so requires a great deal of ingenuity and effort. Fortunately, for many purposes we can derive much valuable information from the use of wavefunctions which at first sight appear to be quite crude.

3.8.1 Slater orbitals

In 1930 J.C. Slater proposed a set of rules for taking into account the influence of shielding. The angular wavefunctions derived from the exact solution of the hydrogen atom Schrödinger equation were preserved, but the radial wavefunctions were replaced by a new set. We shall write the Slater radial wavefunctions as $R_S(r)$ where

$$R_S(r) = N_n r^{n-1} e^{-\zeta r/a_0}.$$

N_n is a normalisation constant for each orbital. The zeta-value (ζ) appearing in the exponential is the quantity which takes screening into account. Slater published rules for determining zeta-values but these rules have been declining in use. Zeta-values (table 3.6) are now calculated by the use of high-speed computers.

The hydrogen-like radial wavefunctions of an atom of atomic number Z can be obtained from table 3.4. The major differences between these and the Slater orbitals of table 3.6 are clear. The Slater-type orbitals (STOs) ignore all but the highest powers of r. This means that the STOs are not very good approximations close to the nucleus, but they improve as r increases. To this extent we might expect the STOs to give better predictions about the first ionisation energies of atoms than their X-ray spectra because X-ray spectra involve changes in electron structures amongst the innermost orbitals.

The second obvious difference in the orbitals is the replacement of the exponential factor e^{-Zr/na_0} by the Slater factor $e^{-\zeta r/a_0}$. If an electron is shielded by the other electrons it feels only a fraction of the total charge at the nucleus. The zeta-value reflects this. We shall call Z_{eff} the effective nuclear charge i.e. the portion that the electron effectively feels. Slater's zeta-value is given by

$$\zeta = Z_{eff}/n$$

where n is the principal quantum number. The computer method of calculating zeta-values is the method of the self-consistent field (SCF).

3.8.2 SCF orbitals

A method for calculating the form of atomic orbitals for complex atoms was invented by D.R. Hartree in 1928. His method is outlined in the flow chart of fig. 3.14. The key idea is that each electron is imagined to behave under the influence of the average field of all the others, together with the field of the nucleus. Armed with a computer and the appropriate programs (now widely available) it is a relatively straightforward matter to solve the Schrödinger equation using Hartree's method to produce an approximate wavefunction for each electron. The skill comes in guessing a set of orbitals to start the procedure. If a poor set is chosen, the method can be extremely tedious. If a good guess is made the iterative procedure produces a set of orbitals that become self-consistent fairly quickly.

To outline the method let us suppose we have just four electrons a, b, c, and d. We guess their initial wavefunctions to be $\psi_1(a)$, $\psi_1(b)$, $\psi_1(c)$, $\psi_1(d)$. Now let us single out electron a and calculate the effective field of b, c, and d. After solving the Schrödinger equation by numerical methods we obtain an improved wavefunction for a, say $\psi_2(a)$. Now we choose electron b and calculate the effective field of $\psi_2(a)$, $\psi_1(c)$, and $\psi_1(d)$. Again we solve the Schrödinger equation to obtain an improved wavefunction $\psi_2(b)$ for electron b. And so the procedure goes on until we become content with the level of agreement between the new set of wavefunctions.

There have been several variations on the theme of the SCF method. Especially, a refinement by V. Fock has given rise to Hartree–Fock SCF orbitals. These are widely used because they are designed to include the demands made by the Pauli exclusion principle. The Hartree–Fock orbitals therefore take some account of spin correlation. However, because the SCF method at heart relies on averaging electron repulsions, it does not give a satisfactory account of electron correlation.

Before leaving this section we shall briefly consider how electron correlation can be taken into account. The most direct method of accommodating radial correlation is not to place two electrons in identical orbitals. For example the two electrons of helium would be placed in orbitals whose exponential factors are different: $\psi_s(1) \propto e^{-\zeta_1 r/a_0}$ and $\psi_{1s}(2) \propto e^{-\zeta_2 r/a_0}$. The different zeta-values imply that one electron is shielded more than the other. This means that one electron is on average nearer to the nucleus than the other. Thus they are less likely to be

Fig. 3.14. Partial flow chart for determining SCF orbitals. The cycle of operation is repeated until a set of wavefunctions is obtained which, when put through the cycle again are not significantly changed, i.e. they become self-consistent.

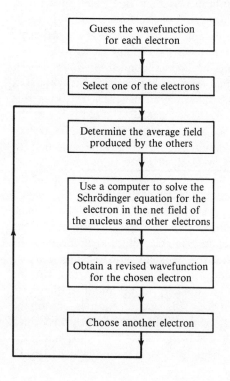

found at the same distance from the nucleus than if they were in identical orbitals.

To take account of angular correlation it is possible to mix orbitals of different symmetries. This has the effect of distorting the electron clouds. For example, if we were to add a degree of p_z orbital symmetry to an s orbital, the electron cloud would no longer be perfectly spherically symmetric. Instead it would be drawn out along the z-axis. The more p_z character we gave the s orbital, the more the distortion would become. This is an application of the idea of overlap of wavefunctions that we met in chapter 2. The overlap of an s and a p_z orbital can be done in two main ways. Either we can form a wavefunction $s + p_z$ or a wavefunction $s - p_z$. The result is two different wavefunctions. If we square the resulting wavefunctions to produce probability density diagrams we find the new wavefunctions have regions of high probability density in the positive and negative z-directions respectively. This is illustrated in fig. 3.15.

Such orbitals are able to satisfy the requirements of minimising angular correlation because the electrons in these orbitals would spend the majority of their time on different sides of the

nucleus. Orbitals which are found by the combination of two or more orbitals of differing symmetries are known as hybrid orbitals. The two hybrid orbitals we have met here are sp hybrids. The importance of hybrid orbitals lies not so much in the formation of orbitals for isolated atoms, but in the formation of orbitals in molecules where their directional properties are very useful in accounting for the shapes of molecules.

Questions

3.11 To normalise the Slater radial wavefunctions we must have

$$N_n^2 \int_0^\infty r^2 |r^{n-1} e^{-\zeta r/a_0}|^2 \, dr = 1.$$

Simplify the integral and use the standard integrals in appendix A to determine the general formula for N_n.

3.12 Using your result from question 3.11, plot graphs of $r^2 |R_s(r)|^2$ for Slater 1s, 2s and 2p orbitals of hydrogen. How do these compare with the true hydrogen atom radial

Fig. 3.15. By combining s and p_z orbitals it is possible to displace charge to lie predominantly along the positive or negative z-axis. That is, the spherical symmetry of the s orbital is destroyed.

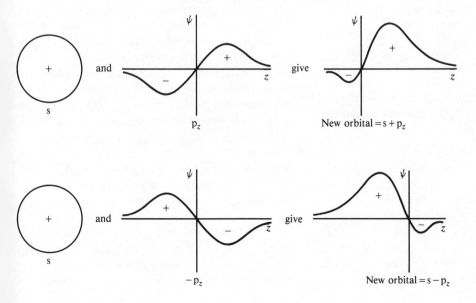

probability density functions of fig. 3.7? (Use a computer for the calculations.)

3.13 If we restrict attention to the z-axis only, then the radius $r = \sqrt{(x^2 + y^2 + z^2)}$ simplifies to $r = \sqrt{z^2}$. (Note that r must always be positive.) Similarly, along the z-axis, $r \cos \theta = z$, which can be positive or negative. Use this information, and a computer to calculate and/or plot a graph of the wavefunction against z for the hybrid orbital $2s + 2p_z$. Repeat the process but this time for $2s + k2p_z$ where k is a small numerical constant, say of the order 0.1 or less. What is a connection between k and angular correlation?

3.9 The total wavefunction of an atom

We now know that, for example, the electron configuration of lithium is $1s^2 2s$ with the implication that the two 1s electrons have opposite spins. The problem is to write down the total wavefunction for the atom. By 'total' we mean the wavefunction that displays both spatial and spin wavefunctions. To solve the problem we must turn to several fundamental ideas in quantum theory. The first is that electrons are indistinguishable – one electron is identical to any other electron. The wavefunction for an atom must take account of this.

If we label the three electrons of lithium as 1, 2, and 3 then we might expect the spatial wavefunction to be $1s(1)1s(2)2s(3)$. However, the principle of indistinguishability tells us that we cannot be sure that the electrons are arranged in this way. The choice $1s(2)1s(3)2s(1)$ is equally likely.

The second fundamental fact of which we must take account is that

the total wavefunction for a system of electrons must be antisymmetric to the exchange of electrons.

Suppose we have two electrons, as in helium. If we write the total wavefunction as $1s(1)\alpha(1)1s(2)\beta(2)$ we have taken account of the fact that both electrons exist in the same orbital, i.e. they have the same spatial wave function, and they have opposite spins. Once again though, we have no reason to think that the choice $1s(1)\beta(1)1s(2)\alpha(2)$ is any the less likely. If we remember that our numbering system is purely artificial we can see that these two choices are

degenerate. As we found in chapter 2, this being the case we can form them into two new combinations by adding or subtracting them. These two choices we shall label $\psi_+(1, 2)$ and $\psi_-(1, 2)$ respectively.

$$\psi_+(1, 2) = 1s(1)\alpha(1)1s(2)\beta(2) \\ + 1s(1)\beta(1)1s(2)\alpha(2)$$

or

$$\psi_+(1, 2) = 1s(1)1s(2)[\alpha(1)\beta(2) + \alpha(2)\beta(1)],$$
$$\psi_-(1, 2) = 1s(1)\alpha(1)1s(2)\beta(2) \\ - 1s(1)\beta(1)1s(2)\alpha(2)$$

or

$$\psi_-(1, 2) = 1s(1)1s(2)[\alpha(1)\beta(2) - \alpha(2)\beta(1)].$$

Now consider what happens when we exchange the two electrons. That is we put electron 2 wherever electron 1 occurs and vice versa. Then

$$\psi_+(2, 1) = 1s(2)1s(1)[\alpha(2)\beta(1) + \alpha(1)\beta(2)] \\ = 1s(1)1s(2)[\alpha(1)\beta(2) + \alpha(2)\beta(1)]$$

so

$$\psi_+(2, 1) = \psi_+(1, 2).$$

Similarly,

$$\psi_-(2, 1) = 1s(2)1s(1)[\alpha(2)\beta(1) - \alpha(1)\beta(2)] \\ = -1s(1)1s(2)[\alpha(1)\beta(2) - \alpha(2)\beta(1)]$$

so

$$\psi_-(2, 1) = -1 \times \psi_-(1, 2).$$

$\psi_+(1, 2)$ is typical of a wavefunction that is symmetric to the exchange of electrons in that it does not change sign. $\psi_-(1, 2)$ is typical of a wavefunction which is antisymmetric to the exchange of electrons. Antisymmetric wavefunctions change sign on the exchange of any two electrons. For electrons, the rule is that the antisymmetric total wavefunction is the correct choice. This is not for any theoretical reason, but because experiments show that this is the case. No one has yet discovered, nor do we believe we are likely to discover, a system of electrons which possesses a symmetric total wavefunction. It is important to realise that on their own spatial wavefunctions and spin wavefunctions can be either symmetric or antisymmetric; but taken together they must form an antisymmetric combination.

For those who know about the properties of determinants it is helpful to recognise that $\psi_-(1, 2)$ can be written in the form of a determinant. See M.3.5. for a short summary of the properties of determinants. In fact,

$$\psi_-(1,2) = \begin{vmatrix} 1s(1)\alpha(1) & 1s(1)\beta(1) \\ 1s(2)\alpha(2) & 1s(2)\beta(2) \end{vmatrix}.$$

The electron number occurs in both the spatial and spin functions so it is just as convenient to put

$$\psi_-(1,2) = \begin{vmatrix} 1s\alpha(1) & 1s\beta(1) \\ 1s\alpha(2) & 1s\beta(2) \end{vmatrix}.$$

If we wish to normalise $\psi_-(1,2)$, as shown in M.3.6, we should put

$$\psi_-(1,2) = \frac{1}{\sqrt{2}} \begin{vmatrix} 1s\alpha(1) & 1s\beta(1) \\ 1s\alpha(2) & 1s\beta(2) \end{vmatrix}.$$

The quantities $1s\alpha(1)$, $1s\beta(1)$, $1s\alpha(2)$, and $1s\beta(2)$ are known as spin-orbitals (note the hyphen). A determinant made up of spin-orbitals is a Slater determinant, named after the same J.C. Slater who invented Slater orbitals. Although we shall not prove it, it can be shown that any antisymmetric wavefunction can be written as a single Slater determinant, or a combination of them.

The virtue of writing wavefunctions in determinantal form can be seen in the following two cases. First, one important property of determinants is that the value of the determinant is unchanged if any multiple of one row, or column, is added to another row, or column. By following M.3.5 we can see that $\psi_-(1,2)$ can be written

$$\psi_-(1,2)$$
$$= -\frac{1}{2\sqrt{2}} \begin{vmatrix} 1s\alpha(1) + 1s\beta(1) & 1s\alpha(1) - 1s\beta(1) \\ 1s\alpha(2) + 1s\beta(2) & 1s\alpha(2) - 1s\beta(2) \end{vmatrix}$$

and this is a perfectly acceptable alternative wavefunction to our original $\psi_-(1,2)$. Indeed, there are an infinite number of alternative ways of writing the total wavefunction for any system of electrons. This fact can be of great use because it means that we are free to choose the wavefunction which best fits our purposes. We shall see in the next chapter that this is particularly helpful when we discuss the bonding in molecules. One combination may be best fitted to explain the spectral properties of a molecule, while another may be best at giving us a picture of the bonds.

Perhaps of more importance is the property of determinants that if any two rows or columns are identical then the determinant is zero. To see the

relevance of this let us try changing the Slater determinant for helium to allow both electrons to have the same spin, α say.

$$\psi_-(1,2) = \frac{1}{\sqrt{2}} \begin{vmatrix} 1s\alpha(1) & 1s\alpha(1) \\ 1s\alpha(2) & 1s\alpha(2) \end{vmatrix}.$$

This determinant is now zero because both columns are the same. Thus we have shown that the determinant will only be non-zero if one of the spins is α and the other β. You should recognise this as one way of stating the Pauli exclusion principle. Indeed, the most powerful and significant way of including the exclusion principle in chemistry is, to re-state the theorem, that

the total wavefunction for a system of electrons must be antisymmetric to the exchange of any pair of electrons.

This is a statement of the Pauli principle. It is of fundamental importance. From it, the exclusion principle can be derived, and much more.

If one of the 1s electrons of helium is promoted to the 2s orbital, the helium atom exists in an excited state with electron configuration 1s2s. It is easy to construct a symmetric spatial wavefunction, $1s(1)2s(2) + 1s(2)2s(1)$, and an antisymmetric spatial wavefunction $1s(1)2s(2) - 1s(2)2s(1)$. As the electrons exist in different orbitals it is possible for the spins to be opposed or parallel. By analogy with the spatial wavefunctions, we can write down symmetric and antisymmetric spin functions as $\alpha(1)\beta(2) + \alpha(2)\beta(1)$ and $\alpha(1)\beta(2) - \alpha(2)\beta(1)$ respectively. When both spins are α, we have $\alpha(1)\alpha(2)$ which is quite clearly symmetric. Similarly, $\beta(1)\beta(2)$ makes another symmetric combination. Applying the Pauli principle, we must combine symmetric spatial wavefunctions with antisymmetric spin wavefunctions, and vice versa. Thus we obtain four possible total wavefunctions as shown in table 3.7.

The spectrum of helium gives us evidence to show that all four possibilities actually can occur.

Although we shall not do so, it is possible to prove that:

   (i) the three symmetric spin wavefunctions each correspond to a total spin of one unit of angular momentum, $\hbar$.

  (ii) the antisymmetric combination gives a total spin angular momentum of zero.

A state with zero spin angular momentum is

Table 3.7. *The four possible total wavefunctions for the excited state 1s2s of helium*

Spatial wavefunction	×	Spin wavefunction
Symmetric	×	Antisymmetric
$\frac{1}{\sqrt{2}}[1s(1)2s(2) + 1s(2)2s(1)]$	×	$\frac{1}{\sqrt{2}}[\alpha(1)\beta(2) - \alpha(2)\beta(1)]$
Antisymmetric	×	Symmetric
$\frac{1}{\sqrt{2}}[1s(1)2s(2) - 1s(2)2s(1)]$	×	$\alpha(1)\alpha(2)$
	or	$\beta(1)\beta(2)$
	or	$\frac{1}{\sqrt{2}}[\alpha(1)\beta(2) + \alpha(2)\beta(1)]$.

The factor $1/\sqrt{2}$ appears in order to normalise the wavefunctions. The method used to carry out the normalisation is shown in M.3.6.

said to be a singlet; that with unit spin angular momentum is a triplet. Alternatively we can say that the spin multiplicity of a singlet state is 1, and of a triplet state 3. If we use the symbol S to stand for the total spin angular momentum in units of $\hbar$, the spin multiplicity is given by

$$\text{multiplicity} = 2S + 1$$

We shall make use of these definitions in later chapters, and in the questions that follow.

Questions

3.14 For a system of N electrons with spin-orbitals $\phi_1, \phi_2, \ldots, \phi_N$ and electrons $1, 2, 3, \ldots, N$ the Slater determinant is

$$\frac{1}{\sqrt{(N!)}} \begin{vmatrix} \phi_1(1) & \phi_1(2) & \phi_1(3) & \cdots & \phi_1(N) \\ \phi_2(1) & \phi_2(2) & \phi_2(3) & \cdots & \phi_2(N) \\ \vdots & \vdots & \vdots & & \vdots \\ \phi_N(1) & \phi_N(2) & \phi_N(3) & \cdots & \phi_N(N) \end{vmatrix}$$

Lithium has the electron configuration $1s^2 2s$. If we assume that the 2s electron has α spin then we can construct three spin-orbitals: $\phi_1 \equiv 1s\alpha$; $\phi_2 \equiv 1s\beta$; $\phi_3 \equiv 2s\alpha$.

(i) Construct the Slater determinant for lithium and multiply it out.
(ii) There is an alternative Slater determinant (total wavefunction) for lithium. What is it?
(iii) What is the spin multiplicity of the $1s^2 2s$ configuration of lithium?

3.15 Prove that the wavefunction $\psi_-(1, 2)$ for helium on p. 47 can be rearranged to give

$$\frac{1}{\sqrt{2}} \begin{vmatrix} 1s\alpha(1) & 1s\alpha(2) \\ 1s\beta(1) & 1s\beta(2) \end{vmatrix}$$

$$= -\frac{1}{2\sqrt{2}} \begin{vmatrix} 1s\alpha(1) + 1s\beta(1) & 1s\alpha(2) + 1s\beta(2) \\ 1s\alpha(1) - 1s\beta(1) & 1s\alpha(2) - 1s\beta(2) \end{vmatrix}$$

You will need to make several changes to the determinant as outlined in M.3.5.

3.16 To make the dependence of the spatial wavefunctions of helium on the positions of the two electrons, r_1, r_2, more explicit, we shall write

$$\psi_+(r_1, r_2) = \frac{1}{\sqrt{2}}[1s(r_1)2s(r_2) + 1s(r_2)2s(r_1)]$$

and

$$\psi_-(r_1, r_2) = \frac{1}{\sqrt{2}}[1s(r_1)2s(r_2) - 1s(r_2)2s(r_1)]$$

for the excited state configuration 1s2s.

(i) Show that $\psi_+(r_1, r_2)$ and $\psi_-(r_1, r_2)$ are orthogonal.
(ii) What happens to each wavefunction and to the corresponding probability density when $r_1 = r_2$?
(iii) Which of the two wavefunctions leads to a lowering of repulsion between the electrons?
(iv) Which function would be associated with a singlet state, and which with a triplet?
(v) What is the reason for the much quoted statement that electrons with parallel spins tend to avoid one another? Has this behaviour anything to do with the charge of the electrons? What is the 'cause'?
(vi) This question has been about spin correlation. What do you think spin correlation

means? What is the origin of spin correlation?

3.17 Electrons have a spin quantum number of $\frac{1}{2}$. Other particles that have the same spin quantum number are protons and neutrons. Such particles are called fermions. Particles with a spin quantum number of 1 are called bosons. Photons and alpha particles are examples of bosons. For a system of bosons the Pauli principle restricts the total wavefunction to be symmetric to the exchange of any two particles. As a consequence, the exclusion principle does not hold for bosons.

(i) If electrons were bosons, what effect would this have on the electron configurations of the elements? Take the elements hydrogen to oxygen as examples.

(ii) The two nuclei of a hydrogen molecule consist of two protons, 1 and 2. Write down the four possible nuclear wavefunctions for the hydrogen molecule. Can you think of an explanation for the fact that a sample of hydrogen gas consists of two types of molecule, called orthohydrogen and parahydrogen, in the ratio 3:1?

3.10 Summary

In this chapter we have seen how the Schröd-inger equation can be applied to the hydrogen atom to yield solutions which, in almost every respect, allow us to calculate precisely the results obtained from experiments. The solutions were called orbitals and we examined their spatial properties together with their dependence on the three quantum numbers n, l, m_l. We saw that the radial and angular parts of the orbitals allowed us to visualise the variation of the probability densities in space. The influence of spin and the Pauli exclusion principle were seen to be fundamental to our understanding of the electron structures of the elements and their ionisation potentials.

One result of electron correlation was that the Schrödinger equation could not be solved exactly. This resulted in our meeting Slater and SCF orbitals, and appreciating the need to account for electron correlation.

In the last section we saw that the Pauli principle was of great importance in rationalising the way electrons behave in atoms, and that the exclusion principle could be derived from it. We also found that the total wavefunction of an atom could be usefully summarised in the form of a Slater determinant.

We shall now turn to the use to which orbital theory can be put in explaining the formation of chemical bonds.

M.3.1

If we take the term

$$\frac{1}{r^2}\frac{\partial}{\partial r}\left(r^2\frac{\partial \psi(r,\theta,\phi)}{\partial r}\right)$$

which appears in the Schrödinger equation, it is clear that the wavefunction is being differentiated with respect to r only. Thus in this case the parts of $\psi(r,\theta,\phi)$ which depend on θ and ϕ are effectively constant. Similarly the terms involving $\partial/\partial\theta$ and $\partial^2/\partial\phi^2$ act only on the parts of $\psi(r,\theta,\phi)$ which respectively depend on θ and ϕ only. This means that we are quite entitled to split $\psi(r,\theta,\phi)$ into three parts:

(i) $R(r)$ depending on r only,

(ii) $\Theta(\theta)$ depending on θ only,

(iii) $\Phi(\phi)$ depending on ϕ only.

Therefore we shall put

$$\psi(r,\theta,\phi) = R(r)\Theta(\theta)\,\Phi(\phi).$$

M.3.2

The radial wavefunctions are related to the associated Laguerre functions. These complicated functions are generated from the formula for the Laguerre polynomials.

$$L_{n+l}^{2l+1}(\rho) = \sum_{j=0}^{n-l-1} (-1)^{j+1}$$

$$\times \frac{[(n+l)!]^2}{(n-l-1-j)!\,(2l+1+j)!\,j!}\rho^j$$

where, in the context of Schrödinger's equation $\rho = 2r/na_0$ and n, l are the principal and azimuthal quantum numbers. To generate $R_{n,l}(r)$ we use

$$R_{n,l}(r) = -\left\{\left(\frac{2}{na_0}\right)^3 \frac{(n-l-1)!}{2n[(n+l)!]^3}\right\}^{1/2}$$
$$\times e^{-\rho/2}\rho^l L_{n+l}^{2l+1}(\rho).$$

Alternatively,

$$R_{n,l}(r)$$
$$= -\left\{\left(\frac{2}{na_0}\right)^3 \frac{(n-l-1)!}{2n[(n+l)!]^3}\right\}^{1/2} \mathscr{L}_{n+l}^{2l+1}(\rho)$$

where

$$\mathscr{L}_{n+l}^{2l+1} = e^{-\rho/2}\, \rho^l L_{n+l}^{2l+1}(\rho)$$

are the associated Laguerre functions.

For example, for a 2s orbital

$$R_{2,0}(r) = -\left\{\left(\frac{1}{a_0}\right)^3 \frac{1}{4(2)^3}\right\}^{1/2} e^{-\rho/2} \times 1 \times L_2^1(\rho)$$

but

$$L_2^1 = \text{sum of terms with } j = 0 \text{ and } j = 1$$
$$= (-1)^1 \frac{2^2}{1 \times 1 \times 1} 1 + (-1)^2 \frac{2^2}{1 \times 2 \times 1}\rho$$
$$= 2(\rho - 2).$$

Thus

$$R_{2,0}(r) = \left(\frac{1}{a_0}\right)^{3/2} \frac{1}{2^{3/2}}(2 - \rho)e^{-\rho/2}$$
$$= \frac{1}{2\sqrt{2}}\left(\frac{1}{a_0}\right)^{3/2}(2 - r/a_0)e^{-r/2a_0}$$

as given in table 3.4.

Taken together, $\Theta_{l,m_l}(\theta)\Phi_{m_l}(\phi)$ form the spherical harmonics. Explicitly,

$$\Theta_{l,m_l}(\theta) = \frac{(-1)^l}{2^l l!}\left\{\frac{(2l+1)}{2}\frac{(l-|m_l|)!}{(l+|m_l|)!}\right\}^{1/2}$$
$$\times \sin^{|m_l|}\theta \frac{d^{l+|m_l|}(\sin^{2l}\theta)}{(d\cos\theta)^{l+|m_l|}}$$

while

$$\Phi_{m_l}(\phi) = \frac{1}{\sqrt{(2\pi)}}e^{im_l\phi}.$$

For example, a 2p$_z$ orbital has $l = 1$ and $m_l = 0$. Then

$$\Theta_{1,0}(\theta) = -\frac{1}{2}\left(\frac{3}{2}\right)^{1/2}\frac{d(\sin^2\theta)}{d\cos\theta}.$$

Because $\cos^2\theta + \sin^2\theta = 1$, it follows that $\sin^2\theta = 1 - \cos^2\theta$. Thus,

$$\frac{d(\sin^2\theta)}{d\cos\theta} = \frac{d(1 - \cos^2\theta)}{d\cos\theta}$$

which is of the form $(d/dx)(1 - x^2)$. Therefore

$$\frac{d(\sin^2\theta)}{d\cos\theta} = -2\cos\theta.$$

So finally we have

$$\Theta_{1,0}(\theta) = (3/2)^{1/2}\cos\theta$$

or

$$\Theta_{1,0}(\theta) = (\sqrt{6}/2)\cos\theta$$

which also agrees with table 3.4

M.3.3

To find the maxima in the radial probability distributions we use the fact these maxima (or minima) must occur when

$$\frac{d}{dr}[r^2|R(r)|^2] = 0.$$

For the 1s orbital we have

$$\frac{d}{dr}\left\{r^2\left[2\left(\frac{1}{a_0}\right)^{3/2}e^{-r/a_0}\right]^2\right\}$$
$$= \frac{4}{a_0^3}\frac{d}{dr}r^2 e^{-2r/a_0}$$
$$= \frac{4}{a_0^3}\left[2r e^{-2r/a_0} - \frac{2r^2}{a_0}e^{-2r/a_0}\right].$$

This is zero when

$$2r - \frac{2r^2}{a_0} = 0,$$

i.e. when $r = 0$ or $r = a_0$, the latter giving the maximum.

If we follow through the same procedure for the 2s orbital, eventually we find that with $k = r/a_0$,

$$k^3 - 8k^2 + 16k - 8 = 0.$$

It is straightforward to show that $k - 2$ is a factor, this giving a minimum at $r = 2a_0$. Extracting $k - 2$ we get $k^2 - 6k + 4 = 0$, which has the solution

$$k = 3 \pm \sqrt{5}.$$

Hence $r = (3 \pm \sqrt{5})a_0$ gives the two maxima. The 2p orbital maximum is easy to find, and it will be left to you to show that it occurs at $r = 4a_0$.

M.3.4

In classical physics the average value of the distance, x, of a particle from the origin is given by $\int x P(x)\,dx$ where $P(x)$ is the probability density function. We have seen that $|\psi(x)|^2$ plays the part of the probability density function in quantum theory. Therefore we shall write

$$x_{av} = \int x |\psi(x)|^2 \, dx$$

as the average value of x.

In the case of the electron in the hydrogen atom this becomes

$$r_{av} = \int r |\psi(r, \theta, \phi)|^2 \, dv$$

because in a three-dimensional problem we must integrate over all space. Then,

$$r_{av} = \int\int\int r |\psi(r, \theta, \phi)|^2 \, r^2 \, dr \, \sin\theta \, d\theta \, d\phi.$$

For a 1s orbital,

$$\psi(r, \theta, \phi) = \frac{1}{\sqrt{\pi}}\left(\frac{1}{a_0}\right)^{3/2} e^{-r/a_0},$$

so

$$r_{av} = \frac{1}{\pi a_0{}^3} \int_0^\infty r^3 e^{-2r/a_0} \, dr \int_0^\pi \sin\theta \, d\theta \int_0^{2\pi} d\phi$$

$$= \frac{1}{\pi a_0{}^3} \times \frac{3!}{(2/a_0)^4} \times [-\cos\theta]_0^\pi [\phi]_0^{2\pi}$$

$$= \frac{1}{\pi a_0{}^3} \times (6a_0{}^4/16) \times 2 \times 2\pi.$$

Thus,

$$r_{av} = \tfrac{3}{2} a_0.$$

In performing the integration we have made use of the standard integrals in appendix A.

The task of showing that r_{av} for 2s and 2p orbitals is $6a_0$ and $5a_0$ respectively can be performed in a similar way.

M.3.5

Some properties of determinants

(i) The product $ad - bc$ can be written as the determinant $\begin{vmatrix} a & c \\ b & d \end{vmatrix}$. A determinant of greater order can be factorised into combinations of determinants of lower order. For example,

$$\begin{vmatrix} a & d & g \\ b & e & h \\ c & f & i \end{vmatrix} = a \begin{vmatrix} e & h \\ f & i \end{vmatrix} - d \begin{vmatrix} b & h \\ c & i \end{vmatrix} + g \begin{vmatrix} b & e \\ c & f \end{vmatrix}.$$

(ii) We can multiply the elements of a row or column by a common factor, provided we make allowance by including a compensating factor outside the determinant. For example,

$$\begin{vmatrix} a & c \\ b & d \end{vmatrix} = \frac{1}{2}\begin{vmatrix} 2a & 2c \\ b & d \end{vmatrix} \quad \text{or}$$

$$\frac{1}{k}\begin{vmatrix} ka & kc \\ b & d \end{vmatrix} \quad \text{or} \quad \frac{1}{k}\begin{vmatrix} ka & c \\ kb & d \end{vmatrix}.$$

This is easily confirmed by multiplying out; e.g.

$$\frac{1}{k}\begin{vmatrix} ka & kc \\ b & d \end{vmatrix} = \frac{1}{k}[kad - kbc] = ad - bc.$$

(iii) The value of the determinant is not

changed by adding a multiple of any one row to another row, or of any one column to another column. For example

$$\begin{vmatrix} a & c \\ b & d \end{vmatrix} = \begin{vmatrix} a+2c & c \\ b+2d & d \end{vmatrix}$$

$$= (a+2c)d - (b+2d)c$$

$$= ad - bc.$$

In general,

$$\begin{vmatrix} a & c \\ b & d \end{vmatrix} = \begin{vmatrix} a+kc & c \\ b+kd & d \end{vmatrix} = \begin{vmatrix} a+Kb & c+Kd \\ b & d \end{vmatrix}.$$

This too can be verified by multiplying out.

(iv) If any two rows or columns of a determinant are equal to one another, then the determinant is zero. This is easy to show because, in the simplest case,

$$\begin{vmatrix} a & a \\ b & b \end{vmatrix} = ab - ba = 0.$$

For those with some deeper mathematical insight, we should say that this is only true if $ab = ba$. If a and b are numbers this is so, but if they were operators this would not necessarily be true.

(v) Determinants are useful for solving simultaneous equations. We shall make use of the following theorem, which we shall not try to prove:

A set of simultaneous equations

$$c_{11}x_1 + c_{12}x_2 + c_{13}x_3 + \cdots + c_{1n}x_n = 0$$
$$c_{21}x_1 + c_{22}x_2 + c_{23}x_3 + \cdots + c_{2n}x_n = 0$$
$$\vdots$$
$$c_{n1}x_1 + c_{n2}x_2 + c_{n3}x_3 + \cdots + c_{nn}x_n = 0$$

in which $x_1, x_2, \ldots, x_n$ cannot be zero only has solutions provided the determinant of the coefficients is zero, i.e.

$$\begin{vmatrix} c_{11} & c_{12} & c_{13} & \cdots & c_{1n} \\ c_{21} & c_{22} & c_{23} & \cdots & c_{2n} \\ \vdots & \vdots & \vdots & & \vdots \\ c_{n1} & c_{n2} & c_{n3} & & c_{nn} \end{vmatrix} = 0.$$

An example should make this clear. The equations

$$ax + 3y = 0$$
$$2x + by = 0$$

have solutions other than $x = y = 0$ if

$$\begin{vmatrix} a & 3 \\ 2 & b \end{vmatrix} = 0,$$

i.e. if $ab = 6$. Provided this is so we can choose any combination of values of a and b we like.

M.3.6

Normalisation of total wavefunctions

A total wavefunction consists of a spatial part and a spin part. To normalise the total wavefunction, the spatial and spin parts must be separately normalised. In addition, for the 1s2s configuration where there are two electrons, these wavefunctions must be normalised with respect to both electrons. We shall show a small volume element of three-dimensional space in the coordinates of electron 1 as dv_1, and that for electron 2 as dv_2. The corresponding spin coordinates will be ds_1 and ds_2.

First, we shall tackle the spatial wavefunction $1s(1)2s(2) + 1s(2)2s(1)$. We shall multiply it by a factor, N, termed the normalising factor. We require

$$N^2 \int\int |1s(1)2s(2) + 1s(2)2s(1)|^2 \, dv_1 \, dv_2 = 1.$$

There are three terms once the multiplication is done. The first of these is

$$N^2 \int\int |1s(1)2s(2)|^2 \, dv_1 \, dv_2$$

$$= N^2 \int |1s(1)|^2 \, dv_1 \times \int |2s(2)|^2 \, dv_2$$

$$= N^2$$

because the coordinates of the two electrons

are independent of one another. Here we should remember that the 1s and 2s orbitals are already normalised, so each integral is equal to one. By similar reasoning we find that the term involving $|1s(2)2s(1)|^2$ also contributes another N^2.

The cross products like

$$N^2 \int\int 1s(1)2s(2)1s(2)2s(1)\,dv_1\,dv_2$$

are better written

$$N^2 \int 1s(1)2s(1)\,dv_1 \times \int 1s(2)2s(2)\,dv_2.$$

But remember that the 1s and 2s orbitals are orthogonal if they both refer to the same set of coordinates. Thus the cross terms are zero. Collecting these results together we have

$$2N^2 = 1$$

so

$$N = \pm 1/\sqrt{2}.$$

The sign is irrelevant because whether we choose the $+$ or $-$ sign, the probability density will be the same. For simplicity we chose the $+$ sign in table 3.7.

Now for the spin terms: we require that

$$N^2 \int\int |\alpha(1)\,\beta(2) - \alpha(2)\beta(1)|^2\,ds_1\,ds_2 = 1.$$

The same type of argument holds for these integrals. For example,

$$N^2 \int\int |\alpha(1)\beta(2)|^2\,ds_1\,ds_2$$
$$= N^2 \int |\alpha(1)|^2\,ds_1 \times \int |\beta(2)|^2\,ds_2$$
$$= N^2$$

because each of $\alpha(1)$ and $\beta(2)$ are separately normalised. Also, integrals like $\int \alpha^*(1)\beta(1)\,ds_1$ are zero because of the orthogonality of the spin functions. If we count up the terms we again find that $N = \pm 1/\sqrt{2}$.

4

Electrons in molecules

4.1 Introduction

In principle, if we wish to understand the behaviour of electrons in molecules, we should first write down the Schrödinger equation for the molecule and then solve it. As we saw in the case of atoms the first step is fairly straightforward once all the kinetic and potential energy terms are identified. The trouble comes with stage two. Unfortunately Schrödinger's equation is impossible to solve exactly for systems which have two or more nuclei and more than one electron. (To see why, refer to M.4.1.) This rules out all molecules.

The result is that we are forced to adopt approximate methods of solution. In this chapter we shall study some of these methods. To begin with we shall only consider molecules containing two ident-

ical atoms such as H_2, O_2, Cl_2. These are the homonuclear diatomic molecules. The heteronuclear diatomic molecules, such as CO, HCl, NO, will then be considered.

The two main theories of chemical bonding are molecular orbital theory and valence bond theory. We shall deal with molecular orbital theory first.

4.2 Homonuclear diatomic molecules

A simple pictorial view of the bonding in a molecule can be built up in the following way. If two atoms approach one another their orbitals will gradually overlap. The wavefunctions of the separate atoms can then interfere, either constructively or destructively. If constructive interference occurs there will be an increase in probability density in the overlap region. There will be an increased probability of finding an electron in the internuclear region and a chemical bond may have formed. If the wavefunctions interfere destructively there will be a decrease in the probability density between the nuclei. Instead of a bonding orbital, an antibonding orbital results.

Imagine two hydrogen atoms, A and B, becoming sufficiently close that their 1s orbitals overlap (fig. 4.2). We shall label the orbitals $1s_A$ and $1s_B$. Constructive interference is produced by the

Fig. 4.2. Two s orbitals may interfere (a) constructively or (b) destructively.

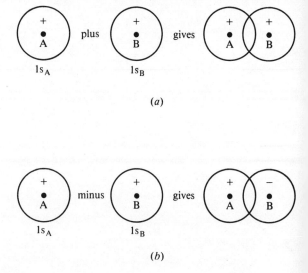

(a)

(b)

Fig. 4.1. The hydrogen molecule, with nuclei A, B, and electrons e_1, e_2 gives rise to six potential energy terms involving $1/r_{A1}$, $1/r_{AB}$ etc.

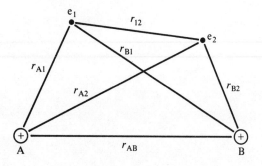

combination $1s_A + 1s_B$. The combination $1s_A - 1s_B$ produces destructive interference. Both the resulting orbitals spread over both atoms. Thus we have formed one bonding molecular orbital and one antiboding molecular orbital.

Notice that the antibonding wavefunction is zero at a point (actually a surface in three dimensions) between the nuclei. This means that the probability density will also be zero there, so in the vicinity there will be a very small probability of finding an electron. Fig. 4.3 illustrates the difference in the probability density distributions for the bonding and antibonding orbitals.

It is convenient to develop a notation for molecular orbitals. If the probability density is symmetrical about the line of centres of the nuclei we have a sigma (σ) bond. If the probability density is not symmetrical in this way, a pi(π) bond results. If the bond is antibonding, we shall use a star on the

Fig. 4.5.(a) The constructive interference of two p_y orbitals gives a $2p_y\pi_u$ bonding orbital. Notice that the orbital has two lobes where the wavefunctions have opposite signs. (b) When destructive interference occurs a $2p_y\pi_g^*$ antibonding orbital results.

Fig. 4.3. Bonding and antibonding orbitals formed by the overlap of two s orbitals. (a) The orbital $1s_A + 1s_B$ is everywhere positive and encompasses regions of space around and between both atoms. The orbital is a σ orbital. (b) The orbital is $1s_A - 1s_B$ and consists of two lobes where the wavefunctions have opposite signs. There is a nodal surface midway between the nuclei. The two lobes form a single σ^* orbital.

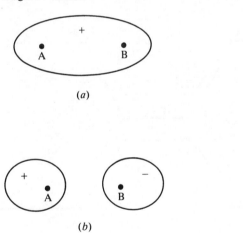

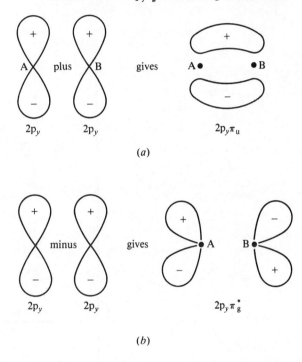

Fig. 4.4. Each of these molecules or ions below has a centre of symmetry, marked with a star.

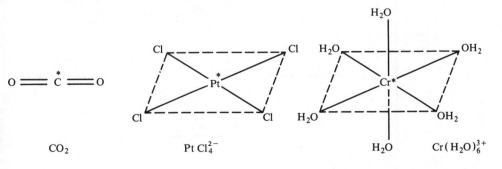

Fig. 4.6. When two orbitals overlap, for example two 1s orbitals, a bonding and an abtibonding orbital result.

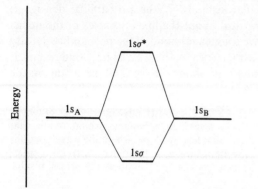

symbol, e.g. σ^* and π^* would be σ and π antibonding orbitals. Here the star has nothing to do with complex conjugation. Subscripts may also be added to show the symmetry properties of the orbitals. If we take a point anywhere in an orbital and find another point diametrically opposed to it through the centre of the molecule, two things may happen. Either the wavefunction at both points will be of the same sign, or it will be of opposite sign. If the former, the wavefunction is an even function, if the latter it is an odd function (see M.4.2.). The orbitals are

Fig. 4.7. Molecular orbital scheme for oxygen and fluorine. The separations between the energy levels are not drawn to scale.

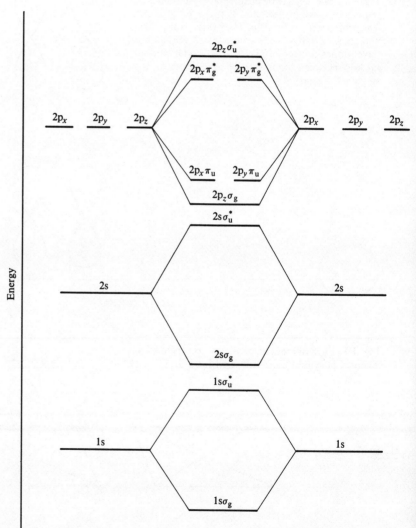

labelled with the latters g and u to show this behaviour after the German words *gerade* and *ungerade* for even and odd.

The bonding orbital formed by the overlap of two 1s orbitals is labelled $1s\sigma_g$; the corresponding antibonding orbital is $1s\sigma_u{}^*$. The simplest π bonds can be formed by the overlap of p orbitals, for example to form $2p\pi_u$ and $2p\pi_g{}^*$ (fig. 4.5). Notice that bonding orbitals are not necessarily even functions.

We are now almost in a position to build up the electron structures of simple molecules by the same type of aufbau process that we used for atoms. The extra piece of information needed is that the formation of a bonding orbital leads to a lowering of energy; that is, stabilisation occurs with respect to the energies of the separate atomic orbitals. An antibonding orbital is associated with a degree of destabilisation. As we shall see later an antibonding orbital is destabilising to a slightly greater extent than a bonding orbital is stabilising (fig. 4.6).

When labelling molecular orbitals we use the convention that the molecular axis is chosen as the z-axis. This is important when labelling σ and π orbitals formed from a set of p orbitals. In fig. 4.7 is shown the general arrangement of the orbital energies for the oxygen and fluorine molecules. Two oxygen atoms contribute 16 electrons all together. We feed the electrons one at a time into the orbitals, always ensuring that the Pauli exclusion principle is obeyed. The first 14 electrons fill the orbitals up to and including the $2p_x\pi_u$ and $2p_y\pi_u$ orbitals. The final two electrons could both enter one of the $2p_x\pi_g{}^*$ or $2p_y\pi_g{}^*$ orbitals, but as expected, electron repulsions are minimised if one electron goes into each of these orbitals. Thus we obtain the electron structure

$$(1s\sigma_g)^2(1s\sigma_u{}^*)^2(2s\sigma_g)^2(2s\sigma_u{}^*)^2(2p_z\sigma_g)^2$$
$$(2p_x\pi_u)^2(2p_y\pi_u)^2(2p_x\pi_g{}^*)^1(2p_y\pi_g{}^*)^1.$$

The arrangement indicates that oxygen should have two unpaired electrons. That is, oxygen should be paramagnetic. This agrees with experiment and therefore it gives some support to our analysis of the bonding in oxygen.

Also we can account for the common description of oxygen as having a double bond. We do this by defining the bond order as bond order = (number of bonding electrons − number of antibonding electrons) × $\frac{1}{2}$.

The arrangement of orbitals is not always the same as that of oxygen or fluorine. The way the orbital energies change is shown on a correlation diagram (fig. 4.8). At one extreme, where the nuclei are infinitely far apart, the molecular orbitals split into distinct atomic orbitals. At the other extreme we have the imaginary case where the two atoms merge into a single whole. This is the united atom limit. The orbitals for the united atom will all have lower energies than in the isolated atoms owing to the doubling of the nuclear charge. Between the separated and united atom limits, the molecular orbital energies will vary in a more or less complicated way.

The main region of interest in the diagram is the point at which the order of energies of the $2p_x\pi_u$, $2p_y\pi_u$ and $2p_z\sigma_g$ orbitals is inverted. The cross-over comes between nitrogen and oxygen. The aufbau method applied to the orbital diagram for nitrogen shows that there are no unpaired electrons, which agrees with the observed diamagnetism of nitrogen.

Fig. 4.8. The correlation diagram shows qualitatively what happens when two single atoms in a homonuclear diatomic molecule merge into one united atom. The molecular orbital diagram for oxygen and fluorine corresponds to the pattern to the right of the vertical line. To the left of the line the pattern gives the molecular orbital diagram for molecules such as nitrogen. Diagram adapted from D.C. Harris & M.D. Bertolucci, Symmetry and Spectroscopy, Oxford University Press, 1978.

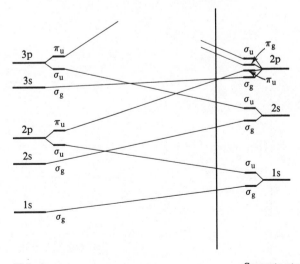

United atom   Separate atoms

Similarly, the predicted bond order is three, once again agreeing with the view of nitrogen as having a triple bond.

Questions

4.1 Using fig. 4.7, write down the electron structure of fluorine, F_2. What is the bond order? Should fluorine be paramagnetic?

4.2 Using fig. 4.8 to adapt fig. 4.7, write down the electron structure of nitrogen, N_2. What is the bond order? Should nitrogen be paramagnetic?

4.3 Repeat questions 4.1 and 4.2 for the ions F_2^+ and N_2^+. Which would you expect to have the stronger bond: (i) F_2 or F_2^+; (ii) N_2 or N_2^+?

4.3 Heteronuclear diatomic molecules

The two atoms in a heteronuclear diatomic molecule will have orbitals of sometimes widely differing energies. This reflects the different degrees to which the nuclei can attract electrons towards them owing to their differing nuclear charges. One main effect is seen in the way the energies of the bonding and antibonding orbitals split (fig. 4.9).

Fig. 4.9. Atom A is more electronegative than atom B. Hence the $1s_A$ orbital is lower in energy than $1s_B$. The molecular orbital diagram is not so symmetrical as in homonuclear diatomic molecules.

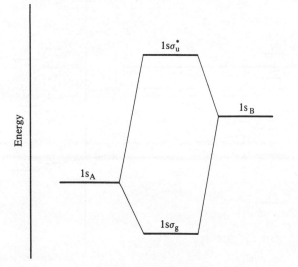

When one of the atoms in a heteronuclear diatomic molecule has a markedly different electronegativity than the other, the molecular orbitals can form a quite different pattern to those that we have met so far. Fluorine, for example, has such a strong attraction between its 1s electrons and the nucleus that these electrons are not used in bonding to the hydrogen atom in hydrogen fluoride. A similar statement is true of the 2s electrons. It is the fluorine 2p orbitals that are closest in energy to the 1s electron of hydrogen. As a consequence, the 1s hydrogen orbital interacts strongly only with the $2p_z$ orbital of fluorine (fig. 4.10). The electron structure is $(1s\sigma)^2(2s\sigma)^2(2p_z\sigma)^2(2p_x\pi)^2(2p_y\pi)^2$. The subscripts g and u do not appear because hydrogen fluoride has no centre of symmetry. The $1s\sigma$, $2s\sigma$, $2p_x\pi$, and $2p_y\pi$ orbitals effectively contain lone pairs of electrons from the fluorine atom.

Questions

4.4 The bonding in the heteronuclear diatomic molecule carbon monoxide is complicated but a bonding π orbital is a main feature of the molecule. Why does the π bond represented in fig. 4.5 not give a realistic impression of the π bond in carbon monoxide? Sketch a better diagram.

4.4 Theoretical background

We have seen that molecular orbital theory can give us some useful insights into the electronic structures of molecules. However, the approach we took was not very sophisticated, and we shall now examine molecular orbital theory in more detail. To begin with we should remember that molecular orbital theory has been developed in its present form only because we cannot exactly solve the Schrödinger equation for molecules. Because we cannot derive the exact molecular wavefunctions or their energies, we assume that molecular orbitals can be formed by taking combinations of the atomic orbitals of the separate atoms. In the parlance of quantum chemistry, we assume that the atomic orbitals form a good basis set for the molecular orbitals. It is also an assumption that we should only take combinations of two orbitals at a time. For example, in oxygen we assumed that the lowest lying σ orbital is formed by a combination of the 1s orbitals on each atom. Although this seems reasonable, we cannot be sure without detailed calculation that the lowest molecular orbital does not have some 2s or 2p character as well.

Take the case of two atoms, A and B, of the same element forming a homopolar diatomic molecule. Let ϕ_A and ϕ_B be two atomic orbitals belonging to A and B respectively. We shall assume that they are normalised:

$$\int |\phi_A|^2 \, d\upsilon = \int |\phi_B|^2 \, d\upsilon = 1.$$

$d\upsilon$ is a small volume element of space, and the integral is taken over all space. If we attempt to form the molecular orbital, ψ, by putting $\psi = \phi_A + \phi_B$ we find in M.4.3 that ψ is not normalised. Properly normalised, it becomes

$$\psi = \frac{1}{\sqrt{(2 + 2S)}}(\phi_A + \phi_B).$$

Similarly, if we assume that $\phi_A - \phi_B$ forms the antibonding orbital ψ^*,

$$\psi^* = \frac{1}{\sqrt{(2 - 2S)}}(\phi_A - \phi_B).$$

The quantity S in these relationships is the overlap integral

$$S = \int \phi_A \phi_B \, d\upsilon$$

which we met in section 2.8.

In M.4.4 we find by some simple algebra that the energies of ψ and ψ^* are

$$E = \frac{\alpha + \beta}{1 + S}$$

$$E^* = \frac{\alpha - \beta}{1 - S}.$$

The symbols α and β, like S, stand for particular types of integral which, once they are evaluated, all

Fig. 4.10. The molecular orbital diagram for hydrogen fluoride.

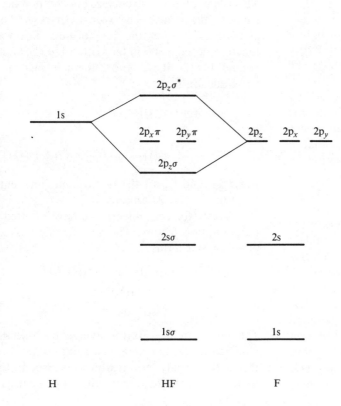

reduce to simple numerical values. The Coulomb integral, α, is given by

$$\alpha = \int \phi_A \hat{H} \phi_A \, dv$$

or

$$\alpha = \int \phi_B \hat{H} \phi_B \, dv$$

when ϕ_A and ϕ_B are identical orbitals. The integral provides a measure of the energy that an electron would have if it existed in an atomic orbital belonging to one atom only but in an environment in which the Hamiltonian is for the molecule as a whole.

The bond integral, β, is given by

$$\beta = \int \phi_B \hat{H} \phi_A \, dv$$

or

$$\beta = \int \phi_A \hat{H} \phi_B \, dv.$$

The value of β is significant only when there is good overlap between orbitals ϕ_A and ϕ_B. If there is no overlap at all then ϕ_A and ϕ_B would not form a bond and the bond integral is zero. On the other hand, if there is good overlap then the contribution of the bond integral will be large. However, notice that in order to bring about a lowering of energy, the value of β must be negative in sign.

The energies of the bonding and antibonding orbitals compared to the energy of the atomic orbitals show that, as we said earlier, an antibonding orbital is somewhat more destabilised than a bonding orbital is stabilised. However, we are no further advanced in explaining why there is a lowering of energy when a bonding molecular orbital is formed. Commonly it is said that the lowering of energy comes about as a result of the lowering of potential energy when there is a build up of electron density between the nuclei. Unfortunately for those who hold this view, detailed calculation shows that the potential energy of electrons in the internuclear region actually increases. On reflection this should make sense because the potential energy of an electron is lowered the nearer it can get to a nucleus. What appears to happen is this. Although an electron in a bonding orbital may have its potential energy increased, because its wavefunc-

tion spreads over a greater distance the curvature of the wavefunction decreases. This means that its kinetic energy will decrease. Such an electron is said to be delocalised, and the attendant drop in energy is the delocalisation energy. At the same time as bond formation occurs there is a general contraction of the orbitals around the nuclei. This brings about an increase in the kinetic energy of the electrons involved but a decrease in their potential energy. It so happens that the two factors tending to lower the energy overcome the two factors tending to raise it. Thus, overall bond formation occurs with a net lowering of energy.

The main problem with the molecular orbital theory we have considered so far is that its predictions for the energies of orbitals are almost invariably wrong. This is shown in the case of hydrogen where not only is the calculated dissociation energy only about 60% of the experimental value, but the equilibrium bond length is predicted to be about 0.01 nm too long. To understand why these errors come about we need to examine the overall molecular orbital for the two electrons in a hydrogen molecule. We shall label the electrons as 1 and 2. This is a matter of convenience only; in practice we cannot distinguish between electrons. Both electrons will enter the $1s\sigma_g$ orbital. The wavefunction of the first will be written $1s\sigma_g(1)$, and the second $1s\sigma_g(2)$. If we ignore the spin factors, the wavefunction is

$$\psi(1,2) = 1s\sigma_g(1)1s\sigma_g(2)$$

or

$$\psi(1,2) = [1s_A(1) + 1s_B(1)][1s_A(2) + 1s_B(2)]$$

where $1s_A$ and $1s_B$ are the 1s wavefunctions centred on atoms A and B respectively.

Again for convenience, we have omitted the normalisation factors. The wavefunction is composed of four terms

$$\begin{aligned}\psi(1,2) &= 1s_A(1)1s_B(2) + 1s_A(2)1s_B(1) \\ &\quad + 1s_A(1)1s_A(2) \\ &\quad + 1s_B(1)1s_B(2).\end{aligned}$$

The third term, $1s_A(1)1s_A(2)$, involves both electrons confined to the 1s orbital belonging to atom A. If this were the only term it would correspond to an ionic structure for hydrogen, with atom A a negative ion and atom B a positive ion. Similarly, $1s_B(1)1s_B(2)$ corresponds to an ionic term but with atom A positive and atom B negative. The first two terms correspond to covalent structures with the electrons shared equally between the atoms. Each of the four terms is given equal prominence in the wavefunction. Thus, the simple molecular orbital theory says that hydrogen is 50% covalent and 50% ionic. The proportion of ionic character is clearly too high, and therein lies the main reason for the poor result for the dissociation energy and equilibrium bond length.

One way of reducing the ionic character is to reconsider the assumption that the best combination for the molecular orbital is of the form $1s_A + 1s_B$. Perhaps it has some of the character of the antibonding orbital $1s\sigma_u^*$. By adding in a contribution from the antibonding state we can produce the molecular orbital

$$\psi(1,2) = 1s_A(1)1s_B(2) + 1s_A(2)1s_B(1) + k[1s_A(1)1s_B(1) + 1s_A(2)1s_B(2)]$$

where k is a numerical constant whose value we are free to choose in order to make the dissociation energy nearest to the observed value. The terms in brackets are the ionic ones, so if k is made small the proportion of ionic character will be small. By doing this we decrease the degree of electron correlation because we are taking account of the tendency of electrons to keep apart. The best result is found for $k \approx 0.26$ which gives a value of about 86% of the observed dissociation energy.

The method used here where portions of two different varieties of orbital are mixed is known as configuration interaction (CI). This is very often used to improve molecular orbitals but is only one of many techniques. A favourite technique is to use SCF calculations, sometimes using standard Slater-type orbitals, but sometimes using different approximate orbitals, especially designed for the molecule in question. Owing to the ease with which computer calculations can be done with them, Gaussian orbitals are very often used. Gaussian orbitals have a factor of the form e^{-Kr^2} included in them. An exponential term like this can be made to give an orbital profile like that of the Gaussian distribution found in statistics. By changing the value of the constant, K, choosing to centre the Gaussians on

any convenient point, and combining with other Gaussians it becomes possible to obtain very good estimates of the measured properties of molecules.

Questions

4.5 Use the method of M.4.4 to show that the energy of the combination $\phi_A - \phi_B$ is $E^* = (\alpha - \beta)/(1 - S)$.

4.6 In M.4.4 we assumed that A and B were atoms of the same element. Now assume that they are different elements. Let $\alpha_A = \int \phi_A \hat{H} \phi_A \, dv$ and $\alpha_B = \int \phi_B \hat{H} \phi_B \, dv$. What are the expressions for the energies of the combinations $\phi_A + \phi_B$ and $\phi_A - \phi_B$?

4.5 Total wavefunctions for molecules

We can write the ground state wavefunction for a hydrogen molecule as $\psi(1,2) = [1s_A(1) + 1s_B(1)][1s_A(2) + 1s_B(2)]$. In this case $\psi(1,2) = \psi(2,1)$ so the wavefunction is symmetric to the exchange of electrons. According to the Pauli principle the total wavefunction must be obtained by combining $\psi(1,2)$ with an antisymmetric spin factor. We shall write $\psi(1,2)$ as $\psi(1,2) = 1s\sigma_g(1)1s\sigma_g(2)$, in which case the total wavefunction becomes

$$\psi_T = 1s\sigma_g(1)1s\sigma_g(2) \frac{1}{\sqrt{2}} [\alpha(1)\beta(2) - \alpha(2)\beta(1)].$$

In determinantal form this is

$$\psi_T = \frac{1}{\sqrt{2}} \begin{vmatrix} 1s\sigma_g\alpha(1) & 1s\sigma_g\alpha(2) \\ 1s\sigma_g\beta(2) & 1s\sigma_g\beta(2) \end{vmatrix}.$$

For molecules with larger numbers of electrons, the determinantal wavefunction can be formed in similar fashion, although the determinant becomes large and inconvenient to write down. For this reason, the determinant is shown in shorthand by merely giving the terms that lie on its diagonal. Also, the normalisation constant is omitted. Thus for hydrogen, ψ is written as

$$\psi_T = |1s\sigma_g\alpha \ 1s\sigma_g\beta|.$$

By writing down the electron structure of a molecule it is then very easy to produce the correct antisymmetrical total wavefunction. This is so even in the case of non-diatomic molecules such as water. We shall not go into detail concerning the method by which the molecular orbitals of water are obtained. Suffice it to say that the task is to combine the two separate 1s orbitals on the two hydrogen atoms with the appropriate orbitals belonging to the oxygen atom. The result is that there are five bonding molecular orbitals which we shall represent by the symbols $1a_1$, $2a_1$, $3a_1$, $1b_2$, and $1b_1$. These symbols arise from group theory and provide information about the symmetry properties of the orbitals. The probability density diagrams for these orbitals are shown in fig. 4.11. The total wavefunction is formed from the determinant

$$|1a_1\alpha \ 1a_1\beta \ 2a_1\alpha \ 2a_1\beta \ 3a_1\alpha \ 3a_1\beta \ 1b_2\alpha \ 1b_2\beta \\ 1b_1\alpha \ 1b_1\beta|$$

which contains ten different types of spin-orbital as we have to accommodate ten electrons in the molecule.

The interesting thing about the orbital diagrams is that they appear to bear very little relation

Fig. 4.11. Probability density diagrams for H_2O. The missing $1a_1$ orbital is the 1s orbital of oxygen. The $1b_1$ orbital is an oxygen p orbital. The $1b_1$ orbital is viewed in the plane of the molecule; the others are viewed looking down on the plane of the molecule.

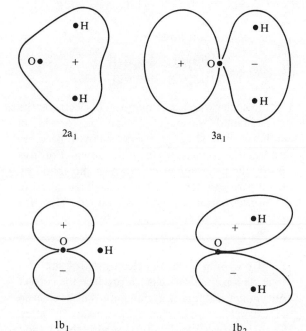

to the ideas that chemists traditionally hold about the bonding in the water molecule. For example it is not possible to identify orbitals that correspond to the notion of water possessing two lone pairs of electrons. Similarly it is hard to relate orbital $2a_1$ with the idea of each oxygen to hydrogen bond being a unique, easily identified part of the molecule. On the molecular orbital scheme it is impossible to view the electrons as localised in individual bonds. By definition, the molecular orbital method envisages electrons to be delocalised over the molecule. However, if we return to section 3.9, there we found that there were an infinite number of different alternative ways of forming the determinant for a wavefunction. This property of determinantal wavefunctions means that in all but a few special cases, such as benzene, it is possible to form localised orbitals out of molecular orbitals. Fairly respectable orbitals localised along the oxygen–hydrogen bonds can be formed by combining $2a_1$, $3a_1$ and $1b_2$ orbitals, and two orbitals corresponding to lone pairs are produced from $2a_1$, $3a_1$, and $1b_1$ orbitals.

The moral of this is that orbitals are not properties of molecules. They are mathematical functions that allow us to calculate and predict the properties. It would be a mistake to think that an atom or molecule 'has' a wavefunction whose form we are trying to discover. Quantum theory allows us to produce as many combinations of wavefunctions as we want in order to help us explain the properties of matter. Molecular orbitals are very good for explaining the spectral properties of molecules; most people find that localised orbitals are better for building mental pictures of bonding in molecules.

4.6 Hückel theory

When we formed bonding and antibonding molecular orbitals by the overlap of two atomic orbitals, ϕ_A and ϕ_B, we had to take account of the overlap between the orbitals. Hückel invented a method of building up molecular orbitals for hydrocarbons by assuming that overlap could be ignored. Strictly this is indefensible, but it makes life much easier and the results it gives are often surprisingly good.

We form a molecular orbital, ψ, by taking the linear combination

$$\psi = c_A\phi_A + c_B\phi_B.$$

As we require ψ to be normalised we must have

$$\int |c_A\phi_A + c_B\phi_B|^2 \, \mathrm{d}v = 1$$

i.e.

$$c_A{}^2 \int |\phi_A|^2 \, \mathrm{d}v + 2c_Ac_B \int \phi_A\phi_B \, \mathrm{d}v$$

$$+ c_B{}^2 \int |\psi_B|^2 \, \mathrm{d}v = 1.$$

But if the overlap integral is zero, and ϕ_A, ϕ_B are independently normalised, this simplifies to

$$c_A{}^2 + c_B{}^2 = 1.$$

Assuming ϕ_A and ϕ_B contribute equally to the molecular orbital means that $c_A{}^2 = c_B{}^2$ so

$$c_A = \pm c_B$$

and

$$c_A = \pm 1/\sqrt{2}.$$

As usual, for convenience we shall take the positive sign. Therefore,

$$\psi = \frac{1}{\sqrt{2}}(\phi_A + \phi_B).$$

Similarly, the antibonding orbital is

$$\psi^* = \frac{1}{\sqrt{2}}(\phi_A - \phi_B).$$

By following through the method of M.4.4 with $S = 0$, it is straightforward to obtain the energies of the bonding and antibonding orbitals:

$$E = \alpha + \beta$$
$$E^* = \alpha - \beta.$$

Because the value of the bond integral, β, is negative, E is lower than E^*, which is what we should expect.

Now we shall apply the method to the types of molecule for which it was intended: hydrocarbons. We shall assume that the σ orbitals of the hydrocarbon chain are present and that a p orbital is left available for bonding on each carbon atom. In the simplest case of ethene, C_2H_4, the results are precisely of the same form as those we have just obtained (fig. 4.12). The resulting orbitals are the well-known π and π^* orbitals. Now let us consider the case of butadiene. The mathematics is a little more complicated. The details can be found in M.4.5. However, we should note here that Hückel theory goes further than just ignoring overlap. It also ignores any interactions between non-neighbouring atoms, and it assumes that all the Coulomb integrals are of equal value, α, and all the bond integrals are of equal value, β. For buta-1,3-diene we obtain the four energy levels given by

$$\alpha + 1.6\beta, \; \alpha + 0.6\beta, \; \alpha - 0.6\beta, \; \alpha - 1.6\beta.$$

Fig. 4.12. The two p orbitals in ethene can combine to form bonding and antibonding π orbitals.

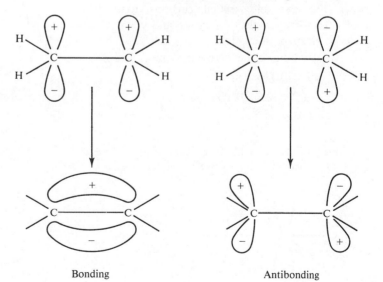

Bonding   Antibonding

The calculation of the four coefficients in the four molecular orbitals is straightforward but tedious. An impression of the shapes of the orbitals is given in fig. 4.14. The coefficients, when squared, give the weightings of the individual contributions to the overall probability density. It follows that a relatively large coefficient, c, at a particular atom implies that c^2 will be relatively large and that an electron is more likely to be found in the region of that atom. This means that the charge density at the atom will also be large, the charge density being proportional to c^2. In the case of buta-1,3-diene there will be two electrons in the lowest lying orbital ψ_1, and two in ψ_2. If we consider the first carbon atom only then the contributions to the charge density on that atom owing to the π orbitals is $(0.37)^2$ from each electron in ψ_1, and $(0.60)^2$ from each electron in ψ_2.

Total charge density on atom 1

$$= 2 \times (0.37)^2 + 2 \times (0.6)^2$$

$$= 1.$$

A quick look at the equations for ψ_1 and ψ_2 shows that this pattern is repeated for each of the four carbon atoms in buta-1,3-diene.

Another piece of information that is sometimes useful to calculate is the π bond order between two atoms. This is obtained by multiplying the two coefficients on the neighbouring atoms and adding together the contribution for each electron. For example, in buta-1,3-diene the π bond order between the first and second carbon atoms is 0.37×0.6 for each electron in ψ_1 and 0.6×0.37 for each electron in ψ_2. Thus,

total π bond order between atoms 1 and 2

$$= 2 \times (0.37 \times 0.6)$$

$$+ 2 \times (0.6 \times 0.37)$$

$$= 0.89.$$

Fig. 4.13. In buta-1,3-diene, $CH_2{=}CH{-}CH{=}CH_2$, there are four p orbitals available to form π bonds.

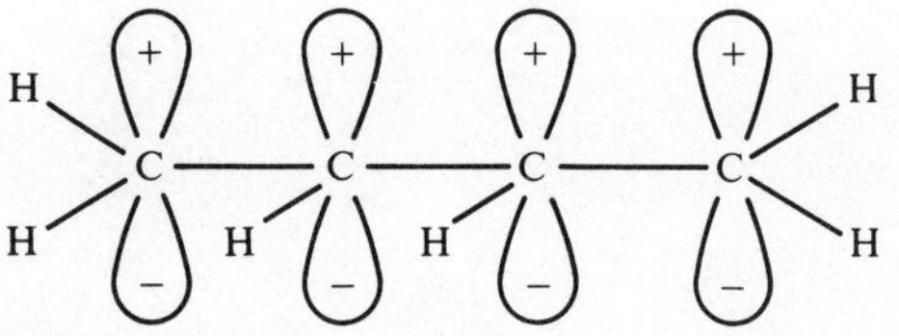

Similar calculation shows that the π bond orders between the third and fourth atoms is also 0.89, while that between the second and third atoms is 0.45. These results are very much in line with the localised bond picture of buta-1,3-diene having two individual π bonds, as in ethene, at each end, with a single bond in the middle of the molecule i.e. $CH_2{=}CH{-}CH{=}CH_2$. There are generalised formulae that can be used to calculate charge densities and bond orders. These are given in M.4.6.

Questions

4.7 The π orbital energies for buta-1,3-diene were

Fig. 4.14. The four π wavefunctions of buta-1,3-diene show a trend from completely bonding, ψ_1, to completely antibonding, ψ_4. The size of the lobes is proportional to the square of the coefficient, c, at each atom.

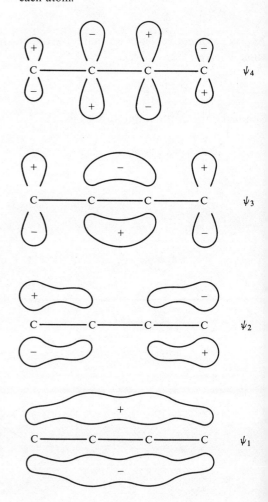

derived in M.4.5, but the calculation of the coefficients was not done. This question should lead you to the values used in section 4.6.

(i) If $\psi = c_1\phi_1 + c_2\phi_2 + c_3\phi_3 + c_4\phi_4$ is to be normalised, $\int |\psi|^2 \, d\upsilon = 1$. Use this to find a relation that must be satisfied by the squares of the coefficients.

(ii) For each value of x, use the set of four simultaneous equations involving x and c_1, c_2, c_3 and c_4 to find the values of the coefficients.

4.8 Use the method of M.4.6 to create, and solve, the secular equation for ethene. Draw an energy level diagram and sketch the wavefunction corresponding to each energy.

4.9 One of the great successes of Hückel theory is the account it gives of the π bonding in benzene. We can write the linear combination of atomic orbitals as $\psi = c_1\phi_1 + c_2\phi_2 + c_3\phi_3 + c_4\phi_4 + c_5\phi_5 + c_6\phi_6$ where each of the ψs is a carbon p orbital.

(i) Follow the method of M.4.6 to form a six by six determinant. Be careful to take account of the fact that atoms 1 and 6 are neighbours.

(ii) By setting the determinant equal to zero and multiplying out the terms, show that the resulting equation to be solved

$$x^6 - 6x^4 + 9x^2 - 4 = 0$$

where $x = (\alpha - E)/\beta$.

Four solutions are $x = \pm 1$ which occur twice. This means that the equation can be factorised with $x^2 - 1$ occurring twice. What is the third factor? What is the third pair of solutions for x?

(iii) Sketch the energy level diagram.

(iv) Place the six electrons on the diagram.

(v) Sketch the forms of the three bonding molecular orbitals:

$$\psi_1 = \frac{1}{\sqrt{6}}(\phi_1 + \phi_2 + \phi_3 + \phi_4 + \phi_5 + \phi_6)$$

$$\psi_2 = \frac{1}{2}(\phi_1 + \phi_2 - \phi_4 - \phi_5)$$

$$\psi_3 = \frac{1}{\sqrt{12}}(\phi_1 + 2\phi_6 + \phi_5 - \phi_2 - 2\phi_3 - \phi_4).$$

4.10 One way of drawing the bonds in benzene is to show it containing three individual π bonds of the type found in ethene.

(i) What is the energy of a π bond in ethene in terms of α and β?

(ii) What would be the total energy of six electrons (i.e. three pairs) in an ethene-like π bond?

(iii) What is the total energy of the six electrons involved in the delocalised π bonds in benzene as predicted from Hückel theory?

(iv) The difference between these two results represents the 'extra' degree of energetic stability of the delocalised electrons in benzene over the electrons in three isolated π bonds. This difference is known as the resonance energy. What value does Hückel theory give for the resonance energy?

4.11 A particularly useful rule is that if a cyclic hydrocarbon with π bonds is to be stable, it should have $4n + 2 \pi$ electrons where n is an integer. For example, benzene has six π electrons, so this fits the rule with $n = 1$. Which of the following species would be expected to be stable:

$$C_3H_3, \ C_3H_3^+, \ C_3H_3^-, \ C_5H_5, \ C_5H_5^+,$$
$$C_5H_5^-.$$

Which of those you have chosen would not be expected to be energetically very stable, in spite of having delocalised electrons? Thus, which might form a stable complex with a transition metal ion?

4.12 Using the forms of the bonding molecular orbitals for benzene as given in question 4.10, calculate the charge density and π bond orders for the six carbon atoms and bonds in benzene. You will not need to calculate all of the coefficients c_1 to c_6 first for ψ_1 to ψ_6. Why not?

4.7 Transition metal complexes

There are two main ways of tackling the problem of explaining the bonding in transition metal complexes. The first, crystal field theory, assumes that the molecules and ions (the ligands) combine with a transition metal to form a set of ionic bonds. The task that crystal field theory sets itself is to explain the way the d orbitals behave in the electrostatic field set up by the ligands. The key result is that the degeneracy of the set of five d orbitals in an isolated atom is removed by the field due to the ligands. The way the degeneracy is lifted is nicely explained by crystal field theory, but we shall say no more about it because the model of the bonding used in the theory is unrealistic. We have good evidence for the fact that d orbitals are used in covalent bonding to ligands, both by way of σ and π bonds. To explain the intricacies of bonding, the alternative theory, ligand field theory, is far better. Ligand field theory is essentially an adaptation of molecular orbital theory and the methods it employs are extensions of those which we have encountered earlier in this chapter.

We begin by considering an octahedral complex and assume that the six ligands will bond to the central transition metal ion through σ orbitals (fig. 4.15). For example, if the ligands were ammonia

Table 4.1. *Combinations of ligand orbitals that form molecular orbitals with the orbitals of a transition metal ion in an octahedral complex.*

Metal orbital	Ligand orbitals
$3d_{z^2}$	$\psi_1 = \dfrac{1}{2\sqrt{3}}(2\phi_5 + 2\phi_6 - \phi_1 - \phi_2 - \phi_3 - \phi_4)$
$3d_{x^2-y^2}$	$\psi_2 = \frac{1}{2}(\phi_1 + \phi_2 - \phi_3 - \phi_4)$
$4s$	$\psi_3 = \dfrac{1}{\sqrt{6}}(\phi_1 + \phi_2 + \phi_3 + \phi_4 + \phi_5 + \phi_6)$
$4p_x$	$\psi_4 = \dfrac{1}{\sqrt{2}}(\phi_1 - \phi_2)$
$4p_y$	$\psi_5 = \dfrac{1}{\sqrt{2}}(\phi_3 - \phi_4)$
$4p_z$	$\psi_6 = \dfrac{1}{\sqrt{2}}(\phi_5 - \phi_6)$

The labels used to describe the molecular orbitals are taken from group theory.

ψ_1 and ψ_2 form the doubly degenerate e_g set
ψ_3 is a single a_{1g} orbital
ψ_4, ψ_5, ψ_6 form a triply degenerate t_{1u} set.

Fig. 4.15. Geometry and numbering system of the σ orbitals for the ligands in an octahedral complex ion.

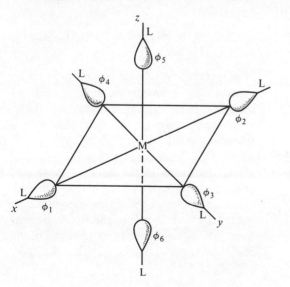

molecules, the lone pair on each nitrogen would be the orbital forming the σ bonds. The problem is to form the correct combinations of the six ligand orbitals which will combine with the available orbitals on the transition metal ion. For a first row transition element, the principal orbitals that may be used are the 4s, 4p and 3d sets. In order to prove the correct form of the combinations of ligand orbitals that are required, it is necessary to take careful account of the symmetries of the transition metal ion orbitals. The details of the calculations would take us too far afield, so the results are given in table 4.1 without proof. The diagrams shown in fig. 4.16 show why the combinations of ligand orbitals do form good choices for overlap with the metal orbitals. Notice that the d_{xy}, d_{yz}, and d_{xz} metal ion orbitals are not used in bonding with the σ orbitals. This set of d orbitals are called the t_{2g} set and are triply degenerate. The d_{z^2} and $d_{x^2-y^2}$ orbitals form the e_g set. These are used in σ bonding. The energy level diagram for an octahedral complex is shown in fig. 4.17.

Fig. 4.16. The origin of the combinations of ligand orbitals to form the molecular orbitals ψ_1 to ψ_6 of Table 4.1.

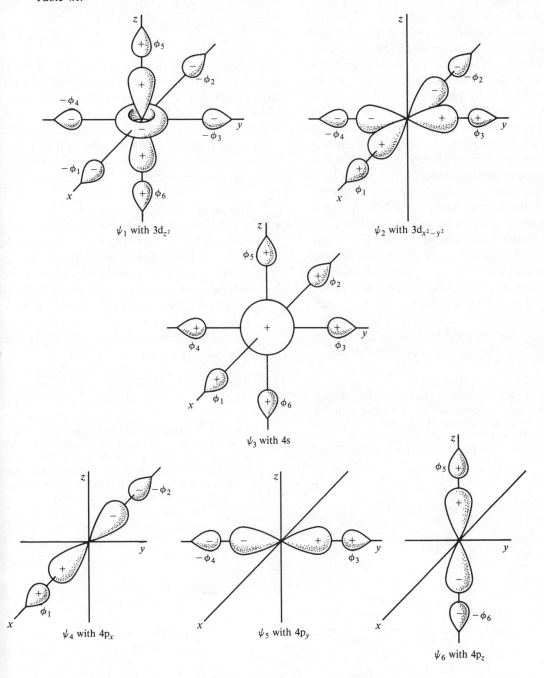

ψ_1 with $3d_{z^2}$

ψ_2 with $3d_{x^2-y^2}$

ψ_3 with 4s

ψ_4 with $4p_x$

ψ_5 with $4p_y$

ψ_6 with $4p_z$

The energy gap between the t_{2g} and e_g^* molecular orbitals, given the symbol Δ or 10Dq, is important. To see why let us investigate the electron structure of the iron (III) complex FeF_6^{3-}. The electron structure of Fe^{3+} is $1s^2 2s^2 2p^6 3s^2 3p^6 3d^5$. Each fluorine will provide two electrons to form the σ bonds, thus we have twelve ligand electrons and five 3d electrons to accommodate. The a_{1g}, t_{1u} and e_g orbitals account for twelve of them, so the problem is to arrange the others between the t_{2g} and e_g^* orbitals. The first three will go into the three separate orbitals forming the t_{2g} set. Where the fourth electron goes is a problem. We know that placing two electrons in the same orbital leads to a raising of energy owing to electron repulsion. If this increase in energy, the pairing energy, is greater than Δ then the electron will go into one of the e_g^* orbitals. If the pairing energy is less than Δ then a lower energy is obtained by two electrons entering the same orbital. Experiment shows that FeF_6^{3-} has a magnetic moment corresponding to five unpaired electrons, thus allowing us to conclude that the pairing energy is greater than Δ in this complex.

However, in $Fe(CN)_6^{3-}$ the position is reversed (fig. 4.18). In this complex the value of Δ is considerably larger than the pairing energy so all five electrons enter the t_{2g} set. The complex has a magnetic moment one-fifth as large as FeF_6^{3-}. As we have seen Δ is influenced by the type of ligand. Indeed it is possible to put ligands into a series, the spectrochemical series, showing the order of increasing effect on the splitting, Δ. A part of the series is

$$I^-, \; Br^-, \; Cl^-, \; F^-, \; H_2O, \; NH_3, \; NO_2^-, \; CN^-.$$

The iodide ion causes the least ligand field splitting, and cyanide ions the greatest. Ligands which cause a large ligand field splitting often do not rely on σ bonding only. Usually they can contribute some π bonding with the t_{2g} orbitals on the metal ion, but the effect on Δ greatly depends on the nature of the π bonding. In some situations it can lead to a decrease rather than an increase in Δ.

Questions

4.13 π bonding in transition metal complexes can come about by the overlap of ligand orbitals with members of the t_{2g} set, i.e. d_{xy}, d_{yz}, or d_{xz} orbitals. Draw a diagram of one of these orbitals and show how overlap could occur with (i) a ligand p orbital, (ii) a ligand d orbital, (iii) a π^* orbital on a ligand, such as ethene, having a π bond of its own. (There are two ways of doing this.)

Fig. 4.17. Energy level diagram for the molecular orbitals in an octahedral transition metal complex. The energy gap between the t_{2g} and e_g^* orbitals is given the symbol Δ or 10Dq in transition metal chemistry. Often this energy gap is called the ligand field splitting.

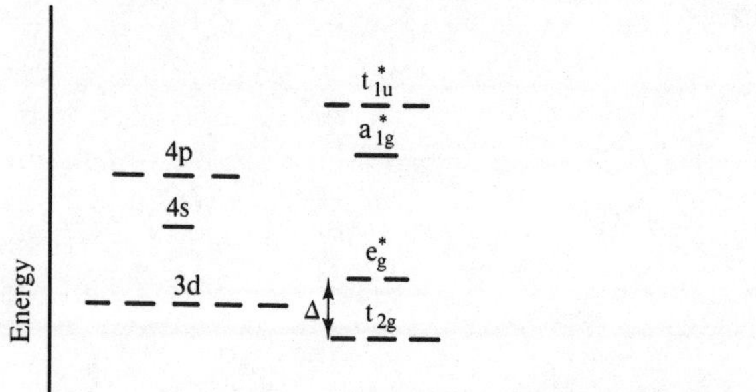

Fig. 4.18. The ligand field splitting in FeF_6^{3-} is less than the pairing energy. This complex, with five unpaired electrons, is a high spin complex. In $Fe(CN)_6^{3-}$ the ligand field splitting is greater than the pairing energy and all five electrons go into the t_{2g} set, thus giving a low spin complex.

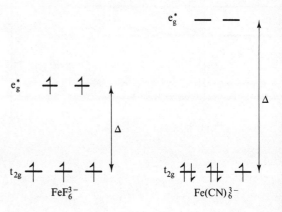

4.14 In a square planar complex, e.g. $PtCl_4^{2-}$, the four ligands are arranged at the corners of a square with the metal ion at the centre. Imagine four ligand σ orbitals, $\phi_1, \phi_2, \phi_3, \phi_4$, to be placed at $+x, -x, +y, -y$ respectively with the metal ion at the origin of the coordinate system. Write down the combination of ligand orbitals that would give a positive overlap with (i) the 4s orbital, (ii) the $3d_{x^2-y^2}$ orbital, (iii) the $4p_x$ orbital, (iv) the $4p_y$ orbital. Which of these would be degenerate?

4.8 Valence bond theory

We have concentrated on molecular orbital theory up to now because for many years it has been the most popular method for quantum chemists to use in calculations. However, as we saw, molecular orbital theory has some drawbacks. Unless corrected it overemphasises the ionic character of bonds and the orbitals it suggests for molecules do not directly correlate with many people's intuitive ideas of localised chemical bonds. In fact, molecular orbital theory was invented after valence bond theory, which, as its name suggests, concentrates our attention on the nature of specific bonds which are formed by the combination of the valence electrons on two neighbouring atoms.

As before we shall begin by looking at the hydrogen molecule. Two hydrogen atoms separated by a large distance will have one electron each in a 1s orbital. Once again labelling the atoms as A and B and the electrons as 1 and 2, the two separated atom wavefunctions will be written $1s_A(1)$ and $1s_B(2)$. The valence bond approach is to imagine the two atoms coming together, complete with their electrons, until their equilibrium distance apart is reached. Then the assumption is that the wavefunction for the molecule is formed by taking the product $1s_A(1)1s_B(2)$ of the two individual wavefunctions. In order to take account of electrons being indistinguishable, the product $1s_A(2)1s_B(1)$ is added. This gives us the unnormalised wavefunction

$$\psi(1,2) = 1s_A(1)1s_B(2) + 1s_A(2)1s_B(1).$$

Clearly this is symmetric to the exchange of electrons and must be combined with the antisymmetric spin wavefunction. The trouble with this wavefunction is the reverse of that with the molecular orbital $[1s_A(1) + 1s_B(1)][1s_A(2) + 1s_B(2)]$. The latter over-estimated the ionic nature of the hydrogen molecule. The valence bond wavefunction over-estimates the covalent nature of the molecule. There are no terms in the valence bond wavefunction which correspond to ionic structures where both electrons may be found centred on the same atom i.e. $1s_A(1)1s_B(1)$ or $1s_A(2)1s_B(2)$. By not allowing the two electrons into the same region of space, valence bond theory overestimates the importance of electron correlation. Molecular orbital theory underestimates it. This inadequacy of valence bond wavefunctions is particularly irksome in the case of bonds formed between atoms of widely differing electronegativities as in HF, HCl, or the hydrogen–oxygen bonds in H_2O. As we might expect the solution is to include a factor which takes account of ionic structures. For example we could write

$$\psi(1,2) = 1s_A(1)1s_B(2) + 1s_A(2)1s_B(1)$$
$$+ \lambda[1s_A(1)1s_B(1) + 1s_A(2)1s_B(2)]$$

where λ is a numerical factor which governs the weighting we might wish to give the ionic structures. Having done this, and comparing with the molecular orbital $\psi(1,2)$ on p. 61, we can see that, after corrections are made, the molecular orbital and valence bond wavefunctions are equivalent to one another.

However, there is one major difference between the two theories. In valence bond theory attention is focussed on the formation of particular, localised, bonds. This is done by imagining the two atoms forming the bond to approach one another with their full complement of electrons. Then we seek the combination of orbitals that best fits our image of a localised bond and most closely fits with experiment. Even in the age of high-speed computers there are formidable problems of calculating the best combination of orbitals, and their energies. This is largely a result of the necessity of taking account of all the possible combinations of the orbitals belonging to one atom with those of the other atom.

In molecular orbital theory we ignore the way the electrons are arranged in the orbitals of the separate atoms. Attention is focussed on forming orbitals that belong to the molecule as a whole – the molecular orbitals. Then, using the aufbau process, we feed electrons into the molecular orbitals, and accept the fact that we lose sight of localised orbitals.

As we have already mentioned, valence bond theory was developed before molecular orbital theory, and it has given rise to several widely used concepts in chemistry. The first of these, resonance, is best considered in the case of benzene. Experiment shows that all the bond lengths in benzene are of the same value and all six bond angles are 120°.

There are two main ways of writing down the bonding between the carbon atoms so as to satisfy the valence of carbon. These are the two Kekulé structures (fig. 4.19). If we ignore the σ framework of the molecule we can write an unnormalised valence bond structure for the π bond between two adjacent atoms as $\phi_1(1)\phi_2(2) + \phi_1(2)\phi_2(1)$ where ϕ_1 is the $2p_z$ orbital on atom 1 and ϕ_2 is the $2p_z$ orbital on atom 2. The overall wavefunction, for the first Kekulé structure shown in fig. 4.19, is a product of three such combinations, i.e.

$$
\begin{aligned}
\psi_1 = &[\phi_1(1)\phi_2(2) + \phi_1(2)\phi_2(1)] \\
&\times [\phi_3(3)\phi_4(4) + \phi_3(4)\phi_4(3)] \\
&\times [\phi_5(5)\phi_6(6) + \phi_5(6)\phi_6(5)].
\end{aligned}
$$

A similar wavefunction, ψ_{II}, can easily be written down for Kekulé structure II. The problem is that neither of these structures, nor their wavefunctions, are alone sufficient to account for the observed properties of benzene. ψ_1, for example, suggests that π bonds are located between atoms 1 and 2, 3 and 4, 5 and 6. These atoms should be linked by double bonds and therefore they should be shorter than the other three bonds. In valence bond theory the difficulty is overcome by assuming the wavefunction which correctly describes the π bonding in benzene is formed by taking a linear combination of ψ_1 and ψ_{II}. The mixing of ψ_1 with ψ_{II} results in a lowering of the energy as compared to the individual Kekulé structures. The lowering is the resonance energy. It is sometimes mistakenly imagined that resonance involves the bonding in benzene oscillating between the two forms I and II. The term resonance should really be interpreted as a means of describing the way the wavefunctions of different valence bond structures are combined to give a resultant that more closely matches the observable properties of a molecule.

The second important concept associated with valence bond theory is that of hybridisation.

Fig. 4.19. Two Kekulé structures for benzene.

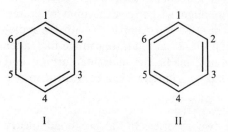

I   II

Fig. 4.20. The set of three unit vectors **i**, **j**, **k** have the same symmetry properties as p_x, p_y, and p_z orbitals. However, note that the diagrams are not to be taken too literally. They are attempts to illustrate both vectors and probability densities on paper. Actually vectors exist in a vector space, and probability densities in what we may call function space. Neither exist in three-dimensional space.

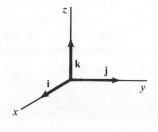

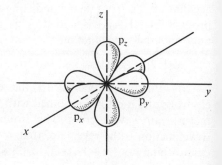

Hybridisation is often invoked when we wish to explain the shapes of molecules in terms of localised bonds. In chapter 3 we saw that it was possible to make an allowance for the angular correlation of electrons in atoms by, for example, adding some p orbital character to an s orbital. The idea is very similar in molecules except that the aim is to produce new orbitals on one atom which will give the right geometry for the molecule and which will overlap strongly with the orbitals on other atoms in the molecule.

The starting point in the theory of hybridisation is the experimentally determined geometry of the molecule. Methane, for example, is a tetrahedral molecule with each H—C—H bond angle being 109°28'. However, the ground state of carbon is $1s^2 2s^2 2p^2$ and at first sight there are insufficient unpaired electrons to give four bonds. This problem is overcome by suggesting that one of the 2s electrons is promoted to the empty member of the 2p set to give the electron configuration $1s^2 2s 2p^3$. Now we have sufficient unpaired electrons (four of them), but without the correct geometry. One way of understanding how the four orbitals can be combined so that their resultants point towards the apices of a tetrahedron is as follows.

p orbitals have definite directional properties with their lobes directed along the *x*-, *y*-, and *z*-axes. In many respects this makes them similar to the unit vectors, **i**, **j**, and **k** that point along the Cartesian axes (fig. 4.20). The scalar products of the unit vectors obey the rules $\mathbf{i} \cdot \mathbf{i} = \mathbf{j} \cdot \mathbf{j} = \mathbf{k} \cdot \mathbf{k} = 1$ and $\mathbf{i} \cdot \mathbf{j} = \mathbf{i} \cdot \mathbf{k} = \mathbf{j} \cdot \mathbf{k} = 0$. Thus, like the p orbitals, the unit vectors are normalised and form an orthogonal set. (The properties of vectors are discussed in more detail in section 6.3.) On the other hand, owing to their spherical symmetry, s orbitals have no preferred directional properties. In this respect they behave like scalars in vector algebra. Thus we can see that the directional properties of the orbital we design must be given by the $2p_x$, $2p_y$, and $2p_z$ orbitals. Fig. 4.21 shows that the apices of a tetrahedron are described by the combination of unit vectors $\mathbf{i} + \mathbf{j} + \mathbf{k}$, $-\mathbf{i} - \mathbf{j} + \mathbf{k}$, $\mathbf{i} - \mathbf{j} - \mathbf{k}$, $-\mathbf{i} + \mathbf{j} - \mathbf{k}$. Then by analogy we expect the four bonds to involve the combinations $2p_x + 2p_y + 2p_z$, $-2p_x - 2p_y + 2p_z$, $2p_x - 2p_y - 2p_z$, $-2p_x + 2p_y - 2p_z$, but the key difference is that we have yet to take the 2s orbital into account. If we remember that each orbital must

be equivalent in all but their directions, it follows that the s orbital must make an equal contribution to each of them. Owing to its spherical symmetry, the simplest way of including the s orbital is to add it to the combinations of p orbitals. This gives, for example, $2s + 2p_x + 2p_y + 2p_z$. In M.4.7 it is shown that each orbital can be normalised to give

$$\psi_1 = \tfrac{1}{2}(2s + 2p_x + 2p_y + 2p_z)$$
$$\psi_2 = \tfrac{1}{2}(2s - 2p_x + 2p_y - 2p_z)$$
$$\psi_3 = \tfrac{1}{2}(2s + 2p_x - 2p_y - 2p_z)$$
$$\psi_4 = \tfrac{1}{2}(2s - 2p_x - 2p_y + 2p_z).$$

ψ_1 to ψ_4 form the set of four tetrahedral hybrid orbitals (fig. 4.22). If we look at the probability density for ψ_1 (M.4.7) we find that the 2p orbitals

Fig. 4.21. The coordinate system for a tetrahedral molecule:

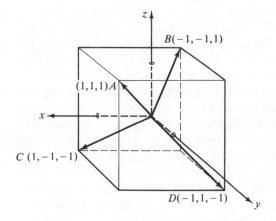

Fig. 4.22. The arrangement of the four sp^3 orbitals for a tetrahedral molecule.

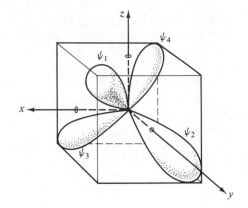

Table 4.2. *Some important hybrid orbital combinations with their corresponding geometries*

Hybrid	Number of hybrid orbitals	Geometry	Example
sp	2	Linear	C_2H_2
sp^2	3	Planar: trigonal	BCl_3
sp^3	4	Tetrahedral	CCl_4
dsp^2 $(s, p_x, p_y, d_{x^2-y^2})$	4	Square planar	$Ni(CN)_4^{2-}$
dsp^3 $(s, p_x, p_y, p_z, d_{z^2})$	5	Trigonal bipyramid	PCl_5
d^2sp^3 $(s, p_x, p_y, p_z, d_{z^2}, d_{x^2-y^2})$	6	Octahedral	$Co(NH_3)_6^{2+}$
d^3s $(s, d_{xy}, d_{xz}, d_{yz})$	4	Tetrahedral	MnO_4^-

provide three times the contribution of the 2s orbital. Such an orbital is termed an sp^3 hybrid.

It is important to understand that the building of the sp^3 hybrids relied on our assuming the promotion of a 2s electron to a 2p orbital. This requires energy to be put into the system. The driving force for the process is the extra energy we obtain by increased overlap of the sp^3 hybrids with the 1s orbitals on the hydrogen atoms in methane. Indeed, the principle of maximum overlap says that the greater the overlap, the stronger the bond.

There is a large number of different hybrid orbitals, each suited to a particular molecular geometry (table 4.2). Those involving d orbitals are important in explaining the geometries of elements such as sulphur and phosphorus as well as in transition metal chemistry (fig. 4.23).

Fig. 4.23. All three structures can be accounted for by the use of the hybrid orbitals in table 4.2.

Questions

4.15 The two Kekulé structures of benzene account for only about 80% of the stabilisation energy. The other 20% can be obtained by including the three Dewar structures (fig. 4.24).

Write down the valence bond wavefunctions for these structures. Call these ψ_{III}, ψ_{IV}, and ψ_{V}. The overall wavefunction for benzene can be formed by taking a linear combination of ψ_I to ψ_V. Now write down (i) the simplest, (ii) the most general, linear combination.

4.16 In addition to sp^3 hybrids, the three sp^2 hybrids and two sp hybrids are often encountered. Assuming the geometry of fig. 4.25.
  (i) write down the positions of the atoms, A, in the first diagram (use the unit vector **i**, **j**, and **k**);
  (ii) translate this into combinations of p_x, p_y, and p_z orbitals;
  (iii) add in the contribution of the 2s orbital;
  (iv) normalise the resulting wavefunctions;
  (v) sketch the shapes of the probability density diagrams for the orbitals.

Fig. 4.24. Three Dewar structures for benzene.

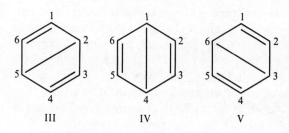

III   IV   V

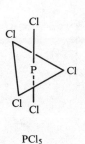

PCl_5

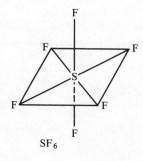

SF_6

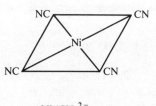

$Ni(CN)_4^{2-}$

You should have now obtained the correct expressions and shapes of the two sp hybrids. Now repeat the process above for the arrangement of the second diagram. This should give you the three sp^2 hybrids.

4.17 Use your results from question 4.16 to sketch the arrangement of σ orbitals in ethyne, C_2H_2, and ethene, C_2H_4. In ethyne there are two p orbitals on each carbon atom not used in σ bonding. Call these the $2p_x$ and $2p_y$ orbitals. In ethene there is one p orbital on each carbon not used in σ bonding. Call these the $2p_y$ orbitals. In each case draw a diagram to show how these orbitals combine to give π bonds in ethyne and ethene.

4.18 Using the wavefunctions for the sp^3 hybrid ψ_1 on p. 71, confirm that it is normalised. Start by working out the terms in $\int |\psi_1|^2 \, d\upsilon$ and make use of the facts that each of the 2s and 2p orbitals is separately normalised and that they form an orthogonal set.

4.19 A perfectly general sp hybrid can be written as a linear combination of an s orbital and a p orbital of the form $s + kp$ where k is a numerical factor. Write such a hybrid as $\psi = N(s + kp)$ where N is a normalisation constant.

Try to write down the value of N by inspection rather than calculation.

4.20 Criticise the following statement:

The only correct way of describing the bonding in methane is to form sp^3 hybrid orbitals. These orbitals predict that the methane molecule should be tetrahedral. This agrees with experiment and gives support to the theory of hybridisation being correct.

4.21 The p orbitals are odd functions, while s orbitals are even functions. We can show this in symbolic form as

$$\left. \begin{array}{l} Ip_x = -p_x \\ Ip_y = -p_y \\ Ip_z = -p_z \end{array} \right\} \quad \text{odd}$$

$$Is = +s \qquad \text{even}$$

where I is the inversion operation which changes $\{x, y, z\}$ to $\{-x, -y, -z\}$.
(i) Are the unit vectors $\mathbf{i}$, $\mathbf{j}$, $\mathbf{k}$ odd or even functions?
(ii) Are the sp^3 hybrid orbitals odd or even functions?

4.9 Summary

Molecules are complicated things; too complicated for Schrödinger's equation to be solved exactly except in a very few cases. We found molecular orbital and valence bond theory to be two complementary, albeit approximate, ways of analysing bonding in molecules. Molecular orbital theory is easier to use in extended computer calculations. It is liked by those who view a molecule as a single, indivisible whole. Valence bond theory, and its offshoot hybridisation, are especially useful when we want to concentrate on a particular part of a molecule. Both theories emphasise the importance of overlap of wavefunctions in the formation of bond orbitals.

We saw that the Pauli principle is of immense significance in explaining the electronic structures of molecules. The use of Slater determinants was found to permit the principle to be applied in a neat way.

Fig. 4.25. The coordinate system used in question 4.16. (a) The geometry for sp hybrids: A—C—A lies along the z-axis. (b) The geometry for sp^2 hybrids: the CA_3 group lies in the xz-plane.

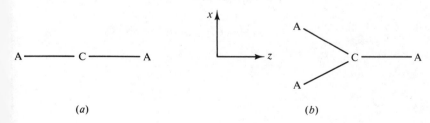

(a)   (b)

M.4.1

The simplest molecule is the hydrogen molecule, H_2, with two protons and two electrons. These give rise to six potential energy terms and, if we ignore the motion of the nuclei, two kinetic energy terms (fig. 4.1). If we label the electrons as 1 and 2, and the protons as A and B, the Schrödinger equation becomes

$$-\frac{\hbar^2}{2m_e}(\nabla_1{}^2 + \nabla_2{}^2)\psi + \frac{e^2}{4\pi\varepsilon_0}\left(\frac{1}{r_{12}} + \frac{1}{r_{AB}}\right.$$

$$\left. -\frac{1}{r_{A1}} - \frac{1}{r_{A2}} - \frac{1}{r_{B1}} - \frac{1}{r_{B2}}\right)\psi = E\psi.$$

If we separate some of the terms in this equation we can put

$$\hat{H}_1\psi + \hat{H}_2\psi + \frac{e^2}{4\pi\varepsilon_0}\left(\frac{1}{r_{12}} + \frac{1}{r_{AB}}\right.$$

$$\left. -\frac{1}{r_{A2}} - \frac{1}{r_{B1}}\right)\psi = E\psi.$$

Here $\hat{H}_1$ and $\hat{H}_2$ are the Hamiltonian operators for separate hydrogen atoms which are of the form

$$\hat{H}_1 = -\frac{\hbar^2}{2m_e}\nabla_1{}^2 - \frac{e^2}{4\pi\varepsilon_0 r_{A1}}.$$

The term in brackets plays the same part as the extra potential energy term which we saw in the Schrödinger equation for a helium atom in section 3.6. It is a term which destroys the symmetry of the potential well. It is this term which makes the equation impossible to solve by exact methods.

M.4.2

Wavefunctions are not the only functions that may be odd or even. In general, a function $f(x, y, z)$ is an even function if $f(x, y, z) = f(-x, -y, -z)$. Simple examples are x^2, $\cos x$, $x^4 y^4 z^4$. A function is an odd function if $f(x, y, z) = -f(-x, -y, -z)$. Examples are y, $\sin x$, $x^3 y^3 z^3$. The change of x, y, z to $-x, -y, -z$ is inversion.

For molecules the idea of inversion involves changing the coordinates of each atom in turn from (x, y, z) to $(-x, -y, -z)$. When this happens, if the atom concerned takes the place of an identical atom then the molecule is said to have a centre of symmetry. Examples of molecules or ions with a centre of symmetry are shown in fig. 4.4.

Take care to note that not all functions are odd or even. For example, the function $f(x) = x^2 + x$ is neither odd nor even because $f(-x) = x^2 - x$. In this case $f(x) \neq f(-x)$ and $f(x) \neq -f(-x)$. Similarly not all molecules have a centre of symmetry. In fact the majority do not. You should have little trouble in thinking of some.

M.4.3

With $\psi = \phi_A + \phi_B$,

$$\int |\psi|^2\, dv = \int |\phi_A + \phi_B|^2\, dv$$

$$= \int |\phi_A|^2\, dv + 2\int \phi_A\phi_B\, dv$$

$$+ \int |\phi_B|^2\, dv$$

$$= 1 + 2S + 1$$

$$= 2 + 2S$$

where S is the overlap integral

$$S = \int \phi_A\phi_B\, dv.$$

If we want ψ to be normalised let us put $\psi = N(\phi_A + \phi_B)$ where N is the normalisation constant. Then

$$\int |\psi|^2\, dv = N^2(2 + 2S).$$

Clearly for $\int |\psi|^2\, dv$ to be unity we must have

$$N^2 = \frac{1}{2 + 2S}$$

so

$$N = \pm \frac{1}{\sqrt{(2 + 2S)}}.$$

By convention it is usual to choose the positive sign and we obtain

$$\psi = \frac{1}{\sqrt{(2 + 2S)}}(\phi_A + \phi_B).$$

M.4.4

We know that the energy of the bonding molecular orbital, ψ, should be given by

$$\hat{H}\psi = E\psi$$

or

$$\hat{H}(\phi_A + \phi_B) = E(\phi_A + \phi_B)$$

where we have cancelled out the normalisation factor which occurs on both sides. Now let us multiply both sides by ϕ_A to give

$$\phi_A\hat{H}\phi_A + \phi_A\hat{H}\phi_B = \phi_A E\phi_A + \phi_A E\phi_B.$$

Remembering that E is merely a number, and that ϕ_A is normalised,

$$\int \phi_A\hat{H}\phi_A \, dv + \int \phi_A\hat{H}\phi_B \, dv$$
$$= E + E\int \phi_A\phi_B \, dv$$

or

$$\alpha + \beta = E + ES$$

where $\alpha = \int \phi_A\hat{H}\phi_A \, dv$ is the Coulomb integral and $\beta = \int \phi_A\hat{H}\phi_B \, dv$ is the bond integral. Thus

$$E = \frac{\alpha + \beta}{1 + S}.$$

Similarly, for the antibonding orbital we can show that $E^* = (\alpha - \beta)/(1 - S)$.

There is an alternative method of obtaining the results for ψ and E which we have given in M.4.3 and M.4.4. The method relies on use of the variation theorem. If we are in the unhappy position of not knowing the true wavefunction for a system (and this is usually the case) we have to make an informed guess of a trial wavefunction. The Variation Theorem says that the energy of the system that we obtain by using the trial wavefunction is always greater or equal to the true energy. The theorem is proved later in section 7.5, as is an example of how it can be applied to a diatomic molecule.

M.4.5

We shall label the four p orbitals in buta-1,3-diene as ϕ_1, ϕ_2, ϕ_3 and ϕ_4 (see fig. 4.13). We know that these orbitals can combine together to form π orbitals. If we call a π orbital ψ then we can put

$$\psi = c_2\phi_1 + c_2\phi_2 + c_3\phi_3 + c_4\phi_4$$

where c_1, c_2, c_3 and c_4 are numerical constants whose values we might hope to determine. If $\hat{H}$ is the Hamiltonian operator and E the energy of the π orbital, we have

$$\hat{H}\psi = E\psi,$$

i.e.

$$\hat{H}(c_1\phi_1 + c_2\phi_2 + c_3\phi_3 + c_4\phi_4)$$
$$= E(c_1\phi_1 + c_2\phi_2 + c_3\phi_3 + c_4\phi_4).$$

We can try to take some progress in the following way: first multiply on the left by ϕ_1 to give

$$\phi_1\hat{H}(c_1\phi_1 + c_2\phi_2 + c_3\phi_3 + c_4\phi_4)$$
$$= E\phi_1(c_1\phi_1 + c_2\phi_2 + c_3\phi_3 + c_4\phi_4).$$

Now if we integrate on both sides we have

$$c_1\int \phi_1\hat{H}\phi_1 \, d_3 + c_2\int \phi_1\hat{H}\phi_2 \, dv$$
$$+ c_3\int \phi_1\hat{H}\phi_3 dv + c_4\int \phi_4\hat{H}\phi_4 \, dv$$
$$= Ec_1\int |\phi_1|^2 \, dv + Ec_2\int \phi_1\phi_2 \, dv$$
$$+ Ec_3\int \phi_1\phi_3 \, dv + Ec_4\int \phi_1\phi_4 \, dv.$$

The integral $\int \phi_1\hat{H}\phi_1 \, dv$ we should recognise as a Coulomb integral, which we shall write α.

$\int \phi_1 \hat{H} \phi_2 \, \mathrm{d}v$ is a bond integral, β, between atoms 1 and 2. The integrals $\int \phi_1 \hat{H} \phi_3 \mathrm{d}v$ and $\int \phi_1 \hat{H} \phi_4 \mathrm{d}v$ are bond integrals between atoms 1 and 3, and 1 and 4 respectively. In Hückel's method we are allowed to assume that the bonding between non-adjacent atoms is so weak that bond integrals like these can be set equal to zero. On the right hand side of the equation, the first integral is unity as the ϕs are normalised. The other three integrals are overlap integrals. Again, in Hückel theory we assume that overlap can be ignored. This may not be unreasonable for integrals involving non-adjacent atoms, but for integrals like $\int \phi_1 \phi_2 \mathrm{d}v$, the approximation is not strictly supportable. However, in practice it works quite well.

Thus we now have

$$c_1 \alpha + c_2 \beta = E c_1$$

or

$$c_1(\alpha - E) + c_2 \beta = 0.$$

By multiplying on the left by ϕ_2, ϕ_3 and ϕ_4 in turn, and integrating we find three further equations:

$$c_1 \beta + c_2(\alpha - E) + c_3 \beta = 0$$
$$c_2 \beta + c_3(\alpha - E) + c_4 \beta = 0$$
$$c_3 \beta + c_4(\alpha - E) = 0.$$

We shall simplify these by dividing through by β and putting $x = (\alpha - E)/\beta$. Then

$$
\begin{aligned}
x c_1 + c_2 & = 0 \\
c_1 + x c_2 + c_3 & = 0 \\
c_2 + x c_3 + c_4 & = 0 \\
c_3 + x c_4 & = 0.
\end{aligned}
$$

This set of four simultaneous equations can be solved by using determinants. From M.3.5, we find that the solutions are given by

$$
\begin{vmatrix}
x & 1 & 0 & 0 \\
1 & x & 1 & 0 \\
0 & 1 & x & 1 \\
0 & 0 & 1 & x
\end{vmatrix} = 0
$$

The determinant is known as the secular determinant for buta-1,3-diene, and the equation is the secular equation.

Multiplying out we find

$$x^4 - 3x^2 + 1 = 0.$$

If we write $y = x^2$,

$$y^2 - 3y + 1 = 0$$

which has the solution

$$y = \frac{3 \pm \sqrt{5}}{2}.$$

Thus

$$x = \pm \sqrt{\left[\frac{(3 \pm \sqrt{5})}{2} \right]} = \pm (\sqrt{5} \pm 1)/2.$$

This means that x can take the values 1.62, 0.62, -0.62, -1.62 so the allowed energies are

$$E_1 = \alpha + 1.62\beta$$
$$E_2 = \alpha + 0.62\beta$$
$$E_3 = \alpha - 0.62\beta$$
$$E_4 = \alpha - 1.62\beta.$$

Note that because β is negative, E_1 is the lowest and E_4 the highest energy. We shall leave it as a task of question 4.7 to show that the values of the coefficients c_1, c_2, c_3, and c_4 which correspond to the given values of x are those which give us the wavefunction:

$\psi_1 = 0.37\phi_1 + 0.60\phi_2 + 0.60\phi_3 + 0.37\phi_4$; energy E_1

$\psi_2 = 0.60\phi_1 + 0.37\phi_2 - 0.37\phi_3 - 0.60\phi_4$; energy E_2

$\psi_3 = 0.60\phi_1 - 0.37\phi_2 - 0.37\phi_3 + 0.60\phi_4$; energy E_3

$\psi_4 = 0.37\phi_1 - 0.60\phi_2 + 0.60\phi_3 - 0.37\phi_4$; energy E_4.

M.4.6

Charge densities can be calculated from the formula

charge density at atom j, $q_j = \sum n_i c_{ij}^{\;2}$

where c_{ij} is the coefficient at atom j for orbital i. The number of electrons in orbital i is n_i. The summation is over the number of electrons in the orbital. Using a similar notation,

bond order between atoms j and k,

$$p_{jk} = \sum n_i c_{ij} c_{ik}.$$

M.4.7

If we call N the normalising factor for ψ_1,

$$\psi_1 = N(2s + 2p_x + 2p_y + 2p_z).$$

Then if we have $\int |\psi_1|^2 dv = 1$,

$$N^2 \int (2s + 2p_x + 2p_y + 2p_z)^2 dv = 1$$

or

$$N^2 \left[\int (2s)^2 dv + \int (2p_x)^2 dv \right.$$
$$\left. + \int (2p_y)^2 dv + \int (2p_z)^2 dv \right.$$

$$\left. + N^2 \left[\text{terms like } \int 2s2p_x \, dv \right. \right.$$

$$\left. \text{and } \int 2p_x 2p_y \, dv \right] = 1.$$

Fortunately, the 2s and 2p orbitals form a normalised and orthogonal set. This means that each integral in the first set of brackets is equal to one, while those in the second set of brackets are all zero. Thus,

$$N^2 \{1 + 3\} = 1 \tag{A}$$

and

$$N = \pm \tfrac{1}{2}.$$

As usual, we choose the positive sign, so

$$\psi_1 = \tfrac{1}{2}(2s + 2p_x + 2p_y + 2p_z).$$

Notice that equation (A) shows that the contribution to the total probability density is in the ratio 1:3 for the 2s and 2p orbitals.

5

Spectroscopy

5.1 Introduction

One of the best ways of investigating the electronic structures of atoms and molecules is to use spectroscopy. The essence of spectroscopy is the observation of the way atoms or molecules exchange energy with the outside world. This exchange may happen when they are heated to high temperatures, as in a simple laboratory flame test; or it may happen in more subtle environments, for example in electric or magnetic fields.

According to Planck's equation energy can be absorbed or emitted in the form of electromagnetic radiation if

$$\Delta E = hf.$$

Here ΔE is the difference in energy between two energy levels, and f is the frequency of the radiation.

If the radiation is emitted then an emission spectrum results; if it is absorbed we observe an absorption spectrum. We have seen that the study of emission spectra of atoms played a large part in the development of quantum theory. This type of spectroscopy is known as electronic spectroscopy owing to the lines occurring when an electron changes its energy from one value to another. Of course molecules can give rise to electronic spectra as well. However, molecules have the added attraction of producing other types of spectra.

A molecule in a gas or liquid will be able to move as a whole from one place to another. That is, it can suffer a translation. Its translational energy,

otherwise known as its kinetic energy, is quantised but the energy levels are remarkably close together. For example, suppose we have a carbon dioxide molecule confined to move to and fro along a tube of length 10^{-2} m. The solution of the particle in a box problem gives its energy levels as

$$E = \frac{n^2 h^2}{8ml^2}.$$

so the difference in energy, ΔE, between two adjacent levels is

$$\Delta E = (2n + 1)\frac{h^2}{8ml^2}.$$

For our carbon dioxide molecule, ΔE is of the order $(2n + 1)10^{-38}$ J. This means that for most purposes the energy levels are so close together as to form a continuous, classical distribution.

The vibrations and rotations of a molecule are another story. Vibrational energy levels and rotational energy levels are definitely not continuous, being separated by amounts that are readily observable. The separations can usually be detected by using infrared and microwave radiation respectively.

In addition to electronic, vibrational, and rotational spectroscopy, three other varieties of spectroscopy are considered in this chapter, all of which are listed in table 5.1. They are photoelectron, electron spin resonance, and nuclear magnetic resonance spectroscopy. The first of these is principally a variety of electronic spectroscopy. The other two are of a rather different type. They rely on the presence of electron and nuclear spins. But more of this later. Right now we shall examine some fundamental ideas that are essential to an understanding of all types of spectroscopy.

First, electromagnetic radiation can be considered to be composed of two oscillating fields. One, the electric field $\mathbf{F}$, the other the magnetic field, $\mathbf{B}$. We can gain a visual impression of the fields by looking at fig. 5.1. Fortunately we can often ignore the fact that the magnitudes of the fields are not constant. This is because the wavelengths of the electromagnetic radiation used in spectroscopy are usually larger than the atom or molecule investigated. Thus, over the length of an atom or molecule the field changes little, as shown in fig. 5.2.

Either, or both, of the fields can bring about a

Table 5.1. *Spectroscopic methods*

Spectroscopic method	Nuclear magnetic resonance	Electron spin resonance	Molecular Rotations	Molecular Vibrations	Electronic	Photoelectron
Region of electromagnetic spectrum	Radio frequency	Microwave	Microwave	Infrared	Visible and ultraviolet	X-ray
Order of magnitude of typical energy change/K J mol^{-1}	10^{-6}	10^{-3}	10^{-1}	10	10^2	10^3

Fig. 5.1. The electric and magnetic field vectors are mutually at right angles. In the diagram, the electric field oscillates in the xz-plane and the magnetic field in the yz-plane.

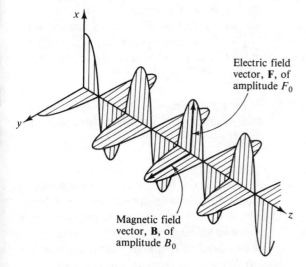

Electric field vector, **F**, of amplitude F_0

Magnetic field vector, **B**, of amplitude B_0

Fig. 5.2. If the wavelength of an electromagnetic wave is sufficiently large, the magnitude of the electric and magnetic fields are approximately constant over the dimensions of the atom or molecule. As the wavelength decreases, i.e. as the energy of the wave increases, this approximation is no longer valid.

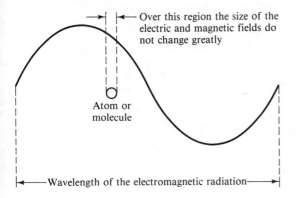

Over this region the size of the electric and magnetic fields do not change greatly

Atom or molecule

←————Wavelength of the electromagnetic radiation————→

transition from one energy level to another. The way the electric field can interact with an atom or molecule is through the presence of an electric dipole moment. This is produced when there is a linear displacement of electric charge. If charges of $+e$ and $-e$ suffer a displacement described by a vector **r**, as shown in fig. 5.3, then the electric dipole moment, **μ**, is given by

$$\boldsymbol{\mu} = -e\mathbf{r}.$$

The energy with which the electric field interacts with the dipole moment is given by the scalar product $-\boldsymbol{\mu}\cdot\mathbf{F}$. The negative signs shows that **μ** and **F** being parallel give the energetically most favourable arrangement, and the antiparallel arrangement the least favourable. Thus far it seems that if there is no dipole moment there can be no interaction involving the electric field. (This is not necessarily true because it is possible to have other types of electric moment, for example electric quadrupole moments; but such niceties need not concern us.) This gives us a clue to solving one of the major puzzles of spectroscopy. Planck's equation says that

Fig. 5.3. The origin of electric and magnetic dipoles. (*a*) An electric dipole moment arises when there is a linear separation of charge. The dipole is given by $\boldsymbol{\mu} = -e\mathbf{r}$. (*b*) A magnetic dipole moment is set up when there is a rotation of charge about an origin. For an electron, the magnetic moment is $\mathbf{m} = -(e/2m_e)\mathbf{l}$ where **l** is the angular momentum vector.

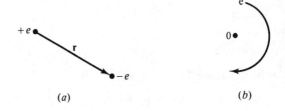

$+e$

r

$-e$

e

0

(*a*)

(*b*)

for every pair of energy levels there is an energy gap such that $\Delta E = hf$. However, it is by no means certain that we will observe radiation of the predicted frequency for every value of ΔE. For example, to a first approximation, transitions between 1s and 2s orbitals do not occur. Such transitions are said to be forbidden. The reason why 1s to 2s transitions do not show up is that both orbitals have spherically symmetric charge distributions. There is no *linear* displacement involved. As a consequence a dipole moment, commonly called a transition dipole moment, cannot occur for a 1s to 2s transition. Hence $-\mu \cdot F$ is zero: there is no energy of interaction with the electric field of the light wave trying to bring about the transition.

Transitions brought about by the electric field are not the only ones that can occur. The magnetic field of electromagnetic radiation can also stir things up. It does so by interacting with the magnetic dipole moment, **m**. The energy of interaction is $-\mathbf{m} \cdot \mathbf{B}$ with **m** given by

$$\mathbf{m} = -(e/2m_e)\mathbf{l}$$

for an electron with angular momentum **l**. Angular momentum can only arise when there is rotational motion, as indicated in fig. 5.3. Even where there is no linear displacement of charge, rotational displacement may occur. Therefore, magnetic dipole transitions may occur even if electric dipole transitions do not. However, magnetic dipole transitions are far less intense than electric dipole transitions and are thus much harder to observe. There are a number of rules – selection rules – which tell us when the different types of transitions can occur. We shall meet various types of selection rule later in this chapter.

Both **μ** and **m** possess interesting mathematical properties which have a bearing on the transitions that are associated with them. Perhaps the most important property is their behaviour with respect to inversion. (See M.4.2 and question 4.21.) Because

$$\mathbf{r} = x\mathbf{i} + y\mathbf{j} + 2\mathbf{k}$$

then

$$I\mu = -\mu.$$

However, because $\mathbf{l} = \mathbf{r} \wedge \mathbf{p}$ and

$$I\mathbf{r} = -\mathbf{r}, \qquad I\mathbf{p} = -\mathbf{p}$$

it follows that

$$Il = +l$$

and

$$Im = +m.$$

This behaviour of **r**, **μ**, **l** and **m** with respect to inversion is summarised by saying that the parity of **r** and **μ** is odd while the parity of **l** and **m** is even. We shall return to this question of parity very soon.

Questions

5.1 The dipole moment is rigorously defined by $\mu = \sum q_i \mathbf{r}_i$ where q_i is the charge (positive or negative) and $\mathbf{r}_i$ is the displacement of the charge from the origin of the coordinate system (fig. 5.4). Note that for a neutral atom, $\sum q_i = 0$. Use this to prove that if the origin is moved by **R** the value of the dipole moment is unchanged.

5.2 In the formula $\mathbf{m} = -(e/2m_e)\mathbf{l}$, what does the minus sign signify about the directions of **m** and **l**?

5.3 The quantity $-e/2m_e$ is known as the magnetogyric ratio. It is the ratio of the magnetic moment to the angular momentum. Classically, it can be derived in the following way. Suppose an electron, charge $-e$, rotates in a circle of radius r with angular frequency $2\pi f$.
  (i) What is the angular momentum?
  (ii) The magnetic moment for a particle of charge q moving in a circular path, sweep-

Fig. 5.4. A collection of charges, q_i, each with a position vector $\mathbf{r}_i$ about some origin (see question 5.1).

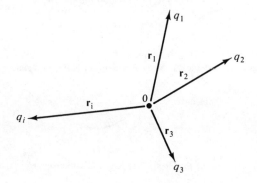

ing out an area A, in time t, is Aq/t. Write this in terms of $-e$, r and f.

(iii) Now work out the magnetogyric ratio.

5.2 Electronic spectroscopy of atoms

Let ψ_1 and ψ_2 be two orbitals of an atom, with corresponding energies E_1 and E_2. If an incoming electromagnetic wave is to excite an electron from ψ_1 to ψ_2, the electric field must bring about a displacement of charge. (We shall ignore the less important influence of the magnetic field.) The displacement must give rise to a dipole moment, however transitory, otherwise it will have no effect. We can put this idea in another way: owing to its occurring during the transition of an electron between two different orbitals, a transition dipole moment must occur. The recipe for calculating the value of the transition dipole moment, $\mathbf{d}$, is

$$\mathbf{d} = \int \psi_2 \hat{\boldsymbol{\mu}} \psi_1 \, \mathrm{d}v.$$

$\hat{\boldsymbol{\mu}}$ is the dipole moment operator. Fortunately this operator does not directly involve partial differentials. In fact it is precisely the same as we defined earlier, i.e. $\hat{\boldsymbol{\mu}} = -e\mathbf{r}$. We can make some useful deductions about $\mathbf{d}$ without a great deal of effort.

First, if we look again at the shapes of the hydrogen atom orbitals it does not take long to see that they are of either odd or even parity. Especially, s and d orbitals are even, and p orbitals are odd. Now suppose ψ_1 is a 1s, ψ_2 is a 2s orbital, and let the electric field of the wave oscillate in the x-direction only. This field can only interact with the x-component of the dipole moment, thus we have

$$d_x = -\int \psi_{2s} e x \psi_{1s} \mathrm{d}v$$

$$= -e \int \psi_{2s} x \psi_{1s} \mathrm{d}v.$$

In terms of parity the integral has the form

$$\psi_{2s} x \psi_{1s} \equiv \text{even} \times \text{odd} \times \text{even}$$
$$\equiv \text{odd}.$$

If we integrate an odd function over all space, for every region where the function is positive there will be an analogous region where it is negative. Therefore, the overall contribution of the positive and negative regions will cancel and the integral will be zero. This can be summarised as follows, albeit in non-rigorous terminology

$$\int (\text{odd function}) \mathrm{d}v = 0.$$

However,

$$\int (\text{even function}) \mathrm{d}v \neq 0.$$

Because the dipole moment is of odd parity we must have either ψ_1 as odd and ψ_2 as even or vice versa. This follows from the facts that

$$\text{even} \times \text{odd} \times \text{odd} = \text{even},$$
$$\text{odd} \times \text{odd} \times \text{even} = \text{even}.$$

We are now in a position to derive our first selection rule, called the Laporte rule which can be summarised in the form

$$\text{g} \leftrightarrow \text{u}, \quad \text{g} \nleftrightarrow \text{g}, \quad \text{u} \nleftrightarrow \text{u}$$

where g and u stand for even and odd parity respectively. The rule says that electric dipole transitions can only occur between orbitals of opposite parity. Especially it means that s$\leftrightarrow$p, and p$\leftrightarrow$d transitions are allowed, but s$\leftrightarrow$s, p$\leftrightarrow$p, and d$\leftrightarrow$d transitions are forbidden. This selection rule correlates with a condition which must be obeyed by the change in angular momentum, Δl, of the electron taking part in the transition, i.e.

$$\Delta l = \pm 1.$$

The selection rule says that, for example, a 1s$\leftrightarrow$2p transition is allowed; but a 1s$\leftrightarrow$3p is forbidden even though 1s$\leftrightarrow$3p is allowed by the Laporte rule. Transitions like 1s$\leftrightarrow$2s are discounted by the Laporte rule and the selection rule on l.

Photons possess an intrinsic spin angular momentum of one unit of angular momentum ($\hbar$). In order to conserve angular momentum in an electric dipole transition if a photon is emitted from an atom, then the atom must lose one unit of angular momentum. Similarly, if a photon is absorbed, the atom must gain one unit of angular momentum. Hence the selection rule $\Delta l = \pm 1$. Incidentally, for other types of transition this selection rule may not hold good owing to the complexities of the changes in angular momentum that can occur, both for photons and for atoms.

Also, we should note that there is a selection rule for the magnetic quantum number, m_l, which says that

$$\Delta m_l = 0, \pm 1$$

but it is important only when an atom is placed in an external electric or magnetic field. For details of the mathematics behind the selection rules, see M.5.1.

There is a further selection rule for the spin,

$$\Delta s = 0.$$

This is to be expected because spin magnetic moments cannot interact with an electric field.

Armed with the selection rules we can account for the lines which appear in the spectra of many atoms, but not all. The selection rules break down when the orbital and spin angular momenta of individual electrons interact strongly with one another. This effect is known as spin–orbit coupling. The heavier the atom, the more important spin–orbit coupling becomes. If we are to understand the basic mechanism of the coupling we must look more closely at the vector properties of orbital and spin angular momentum.

The orbital angular momentum of a body, position vector **r**, moving with momentum **p**, is

$$\mathbf{l} = \mathbf{r} \wedge \mathbf{p}$$

with the direction of **l** being perpendicular to the plane of rotation (fig. 5.5). In classical physics we can make sufficient observations on a rotating body, such as a top, to allow the three components, l_x, l_y, l_z to be determined with arbitrary precision (fig. 5.6). In quantum theory the situation is different. It turns out that on an atomic scale we can measure one

component, l_z say, with precision but if this is done we cannot also measure the other two components, l_x, l_y, with complete precision (fig. 5.7). This is not because of poor experimental method but a result of the inherent nature of the submicroscopic world. The idea that the magnitude of the angular momentum is quantised also applies to l_z. The quantised values of l_z are determined by the magnetic quantum number, m_l, which appears in the solutions of the hydrogen atom problem. The actual magnitude of the angular momentum is given by $\sqrt{[l(l+1)]}\hbar$ where l is the angular momentum quantum number. The z-component has values $m_l\hbar$, or more in keeping

Fig. 5.6. A spinning top will spin on its own axis and precess about the vertical z-axis. It precesses at the Larmor frequency. In classical physics all three components l_x, l_y, and l_z of **l** can be determined without restriction.

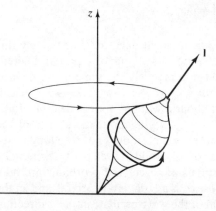

Fig. 5.7. In quantum theory, as **l** precesses about the z-axis, l_z remains constant while l_x and l_y change in a way which we cannot completely determine.

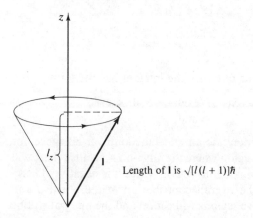

Length of **l** is $\sqrt{[l(l+1)]}\hbar$

Fig. 5.5. The angular momentum vector, **l**, is perpendicular to both the position vector, **r**, and the momentum vector, **p**.

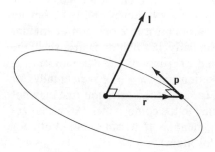

with our present notation, $l_z\hbar$. However, to simplify matters we will not maintain the distinction between the vector **l**, its component l_z, and their quantum numbers. For example, we will speak of an angular momentum of 1 unit with the z-component having allowed values $-1, 0, +1$ when really we should state the magnitudes as $\sqrt{2}\hbar$ and $-\hbar, 0, +\hbar$ respectively. This is general practice and should not cause too much confusion.

We can show the relationship between l and l_z on a diagram as in fig. 5.8. Just as will a classical top, the angular momentum vector will precess about the z-axis producing a cone of orientations. Notice that because we cannot determine l_x and l_y as well as l_z, we cannot be sure of the precise position of the angular momentum vector as it precesses.

Incidentally, we should not forget that atoms know nothing of our coordinate system. The x-, y-, and z-axes are all equivalent until we choose to single out a particular direction in space. This may be done by switching on a magnetic field which defines an axis both for us and for the atom or molecule. Space is no longer isotropic i.e. the same in all directions, it has become anisotropic. Convention has it that the preferred axis is labelled as the z-axis, but this is pure convention; the x- or y-axes would do just as well.

Suppose an atom consists of n electrons. Each electron can be given its own orbital and spin angular momentum vectors, and these vectors combined to form a resultant. In light atoms it turns out that the most sensible way to combine them is for all the orbital momenta to combine to form a resultant amongst themselves, i.e.

$$\mathbf{L} = \mathbf{l}_1 + \mathbf{l}_2 + \mathbf{l}_3 + \cdots + \mathbf{l}_n.$$

Similarly all the spin momenta combine to form their own resultant

$$\mathbf{S} = \mathbf{s}_1 + \mathbf{s}_2 + \mathbf{s}_3 + \cdots + \mathbf{s}_n.$$

Then we choose to form the total angular momentum, **J**, from the resultant of **L** and **S**:

$$\mathbf{J} = \mathbf{L} + \mathbf{S}.$$

J can take the values $L + S, L + S - 1, L + S - 2, \ldots$ to $|L - S|$. Like **L**, J can never be negative, although its projection J_z can be negative. As we might expect, strictly, the values that J can take are quantised according to the rule $\sqrt{[j(j+1)]}\hbar$ and its projections as the z-axis are given, in units of $\hbar$, by j_z ranging through $j, j - 1, j - 2, \ldots, 1, 0, -1, \ldots, -j + 1$, to $-j$. This manner of coupling spin and orbital angular momenta is known as Russell–

Fig. 5.8. The allowed values of l_z for angular momentum of 1 unit (strictly $\sqrt{2}\hbar$) are $+1\hbar, 0$, and $-1\hbar$.

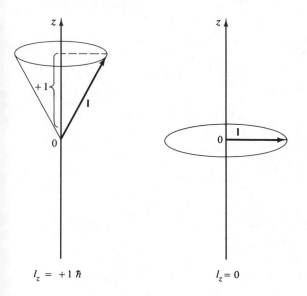
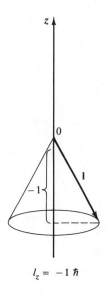

$l_z = +1\,\hbar$ $l_z = 0$ $l_z = -1\,\hbar$

Saunders coupling, and is shown pictorially in fig. 5.9.

Now compare Russell–Saunders coupling to that which occurs to a great extent in heavy atoms, i.e. especially those with atomic numbers greater than 70. For these atoms j–j coupling occurs (fig. 5.10). This form of coupling arises when the orbital and spin momenta of an individual electron interact very strongly together. That is, when there is a strong spin–orbit coupling. It is possible to

Fig. 5.9. In Russell–Saunders coupling the individual orbital angular momenta are coupled together. Then the spin angular momenta are coupled. Finally the resultant orbital and spin angular momentum vectors are coupled to give the total angular moment, **J**.

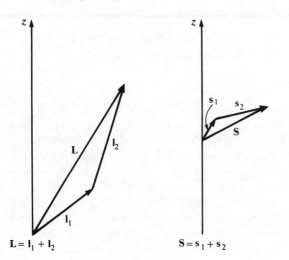

$$\mathbf{L} = \mathbf{l}_1 + \mathbf{l}_2 \qquad \mathbf{S} = \mathbf{s}_1 + \mathbf{s}_2 \qquad \mathbf{J} = \mathbf{L} + \mathbf{S}$$

Fig. 5.10. In j–j coupling the individual orbital and spin angular momenta are coupled together. Then the individual **j**s are coupled to give the total angular momentum, **J**.

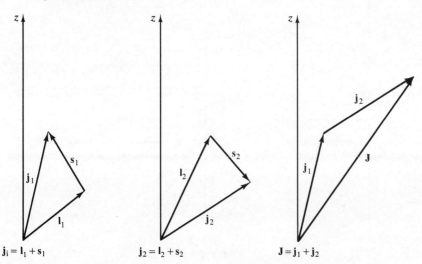

$$\mathbf{j}_1 = \mathbf{l}_1 + \mathbf{s}_1 \qquad \mathbf{j}_2 = \mathbf{l}_2 + \mathbf{s}_2 \qquad \mathbf{J} = \mathbf{j}_1 + \mathbf{j}_2$$

calculate the effect of spin–orbit coupling on the energy of an electron. It is

$$E_{s \cdot o} = \tfrac{1}{2}\zeta_{n,l}[j(j+1) - l(l+1) - s(s+1)]$$

for an electron of principal quantum number n and angular momentum l. As shown in question 5.8, the spin–orbit coupling constant $\zeta_{n,l}$ depends on the fourth power of the atomic number, thus emphasising its importance for heavy atoms. What appears to happen is that the electrons, which are under the influence of a strongly charged nucleus, are drawn closer to the nucleus. Being confined to a small volume of space their wavefunctions increase in curvature and their speed increases. Not only do relativistic effects become important but also the motion of an electron gives rise to a strong magnetic field. This field interacts strongly with the spin magnetic moment. In this way the $\mathbf{l}$ and $\mathbf{s}$ vectors of individual electrons couple together to give a result-ant, $\mathbf{j}$. So,

$$\mathbf{j}_1 = \mathbf{l}_1 + \mathbf{s}_1, \, \mathbf{j}_2 = \mathbf{l}_2 + \mathbf{s}_2, \ldots, \mathbf{j}_n = \mathbf{l}_n + \mathbf{s}_n.$$

The total angular momentum is found by forming the resultant of the $\mathbf{j}$s

$$\mathbf{J} = \mathbf{j}_1 + \mathbf{j}_2 + \mathbf{j}_3 + \cdots + \mathbf{j}_n.$$

Unlike Russell–Saunders coupling, spin–orbit coupling is a property of nature. Nothing in quantum theory *forces* the choice of Russell–Saunders coupling, but spin–orbit coupling *is* forced upon us through the nature of electrons and nuclei. However, in the lighter atoms spin–orbit coupling is insignificant and experience shows that Russell–Saunders coupling is the best choice that we can make. We can see the usefulness of this coupling scheme if we attempt to explain the appearance of atomic spectra. This task is greatly assisted by adopting a special notation. Because of the variety of the different values of L, S and J it has been found useful to introduce *term* symbols which help to define the state in which an atom finds itself. For example, if $L = 0$ and $S = \tfrac{1}{2}$ we will obtain a different state to one with $L = 1$ and $S = \tfrac{1}{2}$.

When the angular momentum, L, is 0, 1, 2, 3, ... the state is labelled as an S, P, D, F, ... state. Notice the similar notation to the labels for orbitals according to their values of l. The *multiplicity* of a state is defined by $2S + 1$, where here S is the symbol for the total spin (not the state symbol). A term

symbol conveys information about the state and its multiplicity. A term symbol has the form ^{2S+1}L. For example when $L = 0$ and $S = \tfrac{1}{2}$ we have the term ^{2}S, read 'doublet-S'; for $L = 1$, $S = \tfrac{1}{2}$ we have ^{2}P, read 'doublet-P'.

It is possible to extend this notation even further by showing the value of J. Then we use a symbol $^{2S+1}L_J$ to define a *level*. For example for $L = 0$, $S = \tfrac{1}{2}$, $J = \tfrac{1}{2}$ we have the level ^{2}S$_{1/2}$ read 'doublet-S-one-half'. There are several ways of finding out the various levels that can be obtained for a given term. If you would like to learn one, try question 5.7.

Once we have produced the terms and levels we can use the selection rules to sort out which can take part in electric dipole transitions.

Electric dipole selection rules

$$\Delta L = 0 \text{ or } \pm 1 \text{ but } 0 \nleftrightarrow 0$$

$$\Delta S = 0$$

$$\Delta J = 0 \text{ or } \pm 1 \text{ but } 0 \nleftrightarrow 0$$

$$g \leftrightarrow u, \, g \nleftrightarrow g, \, u \nleftrightarrow u.$$

In fig. 5.11(a) is shown a diagram, named after its inventor Grotrian, illustrating the origin of the main lines in the spectrum of helium. In its ground state the atom has the configuration $1s^2$ implying

Fig. 5.11. The origin of the spectral lines (*a*) in parahelium, (*b*) in orthohelium.

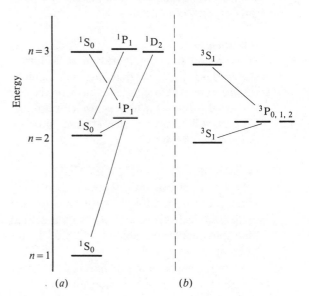

(*a*)   (*b*)

that the electrons have opposed spins. This gives rise to the level $^1S_0$. Owing to the selection rule for L, a transition cannot be made to another S state, but one may be made to a P state. Because $\Delta S = 0$, the spins of the electrons must remain opposed and therefore only singlet P states can be arrived at.

However, the helium spectrum is more complicated than this picture suggests, for mixed in with the lines of fig. 5.11(a) are a completely different set. These occur for transitions between triplet states, i.e. states in which the two electrons have parallel spins, as shown in fig. 5.11(b). When the spectrum of helium was first investigated it was thought that the different series were due to two different varieties of helium, parahelium and orthohelium. However quantum theory shows that the spectrum of parahelium corresponds to transitions between singlet states, and orthohelium to transitions between triplet states.

The appearance of the spectra of a large number of atoms and ions can be explained in much the same way as that of helium. A word of caution though. The selection rules for electric dipole radiation that we have used only hold good for light atoms. For heavy atoms when spin–orbit coupling becomes important different selection rules must be applied. Also, remember that electric dipole forbidden transitions may occur by virtue of magnetic dipole, and other, different types of transition. We shall not, however, delve any deeper into these matters. Instead we shall turn to a description of the electronic spectra of molecules.

Questions

5.4 The ground state configuration of lithium is $1s^2 2s$. Use the fact that the net angular momentum of a closed shell of electrons is zero to help determine the term symbol for lithium. Repeat this task for the ground state of scandium, $[Ar]4s^2 3d$.

5.5 What is the term symbol for the configuration $[Ar]4s^2 3d^2$ of titanium? This is a more difficult problem because now we have to deal with two electrons each with its own orbital and spin angular momentum. First deal with the spins:

(i) are the spins paired or parallel? What is the total spin and the multiplicity?

(ii) as $\mathbf{L} = \mathbf{l}_1 + \mathbf{l}_2$, it follows that $L_z = l_{1z} + l_{2z}$. What are the possible values of l_z for each d orbital? What combinations of l_{1z} and l_{2z} may be formed? Hint: remember Hund's rules and the Pauli exclusion principle.

(iii) What value of L can give rise to the set of L_z values you have found?

(iv) Now write down the term symbol.

5.6 Hund's rules state that for a given electron configuration:

(I) the term with the greatest spin multiplicity has the lowest energy;

(II) if the multiplicity of two terms is the same then that with the greatest orbital angular momentum has the lowest energy;

(III) if both the multiplicity and the orbital angular momentum are the same, then if the electron shell is less than half-filled the lowest energy is given by the lowest value of J. The converse is true if the shell is more than half full.

(i) Apply the first two rules to determine the term symbol (not level) for nickel $[Ar]4s^2 3d^8$. Note that in order to make L a maximum, L_z has to be a maximum.

(ii) A set of ^3P states, $^3P_0$, $^3P_1$, $^3P_2$ can arise for the electron configuration 1s2p. Place these levels in order of increasing energy.

(iii) Another set of levels $^3D_1$, $^3D_2$, $^3D_3$ arise for the configuration 1s3d. Place these in order of increasing energy.

(iv) The ^3D set all lie above the ^3P set in energy. Does this contradict Hund's rules?

(v) Draw an energy level diagram for the ^3P and ^3D levels. The scale does not matter but assume that the splittings between the levels are unequal.

(vi) Use the selection rules to predict the number of lines that might appear in a spectrum owing to transitions between the levels.

5.7 This question illustrates one way of determining the J values, and hence the levels, for a single p electron.

(i) Suppose $J = \frac{3}{2}$, what values of J_z are allowed?

(ii) Likewise, for $J = \frac{1}{2}$ what are the values of J_z?

(iii) If $L = 1$, what values of L_z are allowed?

(iv) What values of S_z are permitted for one electron?

(v) Given that $J_z = L_z + S_z$, list all the allowed values of J_z, even if there are repeated values.

(vi) Look at your answers to (i) and (ii) and say which J values could give rise to these values of J_z.

(vii) Write down the levels.

5.8 The term in the Schrödinger equation which gives rise to spin–orbit coupling is the Hamiltonian operator

$$\hat{H}_{s.o} = \frac{1}{2m_e^{\,2}c^2}\left(\frac{1}{r}\frac{\partial V}{\partial r}\right)\mathbf{l}\cdot\mathbf{s}$$

with V the potential energy of the electron.

When $\hat{H}_{s.o}$ acts on a wavefunction the energy obtained is

$$E_{s.o} = \tfrac{1}{2}\zeta_{n,l}[j(j+1) - l(l+1)$$
$$- s(s+1)] \qquad \text{(A)}$$

with

$$\zeta_{n,l} = \frac{h^2 e^2 Z^4}{32\pi^3\varepsilon_0 m_e^{\,2}c^2 a_0^{\,3} n^3 l(l+\tfrac{1}{2})(l+1)}$$

and $a_0 = \varepsilon_0 h^2/\pi m_e e^2$ is the first Bohr radius.

There is a strong yellow line in the spectrum of sodium which is in fact a doublet. The transition involved is between 3s and 3p orbitals, giving rise to ^{2}S and ^{2}P terms.

(i) What levels are possible for these terms?

(ii) Using equation (A) calculate, in terms of $\zeta_{n,l}$, the spin–orbit energy for each level.

(iii) Draw an energy level diagram showing the transitions causing the doublet.

(iv) The two transitions occur at 589.593 nm and 588.996 nm. Estimate the value of $\zeta_{n,l}$ in cm^{-1}.

5.3 Electronic spectroscopy of molecules

You may remember that in using the Schrödinger equation for the atoms and molecules of chapters 3 and 4, we left out the terms relating to the motion of the nuclei. This is reasonable because the great difference between the masses of electrons and nuclei means that the nuclei move much more slowly than the electrons. Thus without too great a loss of accuracy we can assume that the movement of the nuclei can be ignored. This assumption is called the Born–Oppenheimer approximation which claims that, to a good approximation, the total wavefunction, ψ_T, for a molecule can be separated into three parts: one for the electronic motion, ψ_e, one for the vibrational motion, ψ_v, and one for the rotational motion, ψ_r, i.e.

$$\psi_T = \psi_e \psi_v \psi_r.$$

If we return to the solution of the two- and three-dimensional particle in a box problem of chapter 2, there we found that a wavefunction like ψ_T gives rise to a total energy which is the sum of the energies corresponding to each contributory wavefunction. Here,

$$E_T = E_e + E_v + E_r.$$

Notice that, as we found in the case of the particle in a box, whereas the wavefunctions multiply, the energies add.

As shown in fig. 5.12, the spacing between electronic energy levels is much greater than between vibrational energy levels. Rotational energy levels are very closely spaced together. Transitions can occur between each of these types of energy level, giving rise to electronic, vibrational, and rotational spectra. Often transitions can occur between combinations of these levels, but for the present we shall concentrate on electronic spectra and assume that the various complicating factors do not occur.

The cylindrical symmetry of a diatomic molecule has a very important effect on the angular momentum of the electrons. Unlike the case of atoms, which are spherically symmetric, the total angular momentum, **L**, is not well defined. The component of **L** along the internuclear axis is much more important. The angular momentum is said to be quantised along the axis, and is given the symbol Λ; Λ is built up by combining the angular momenta, λ_i, along the axis of the individual electrons:

$$\Lambda = \sum \lambda_i.$$

Λ can take the integral values $0, 1, 2, \ldots$ which define the states Σ, Π, Δ, Φ,... which are analogous to the atomic states S, P, D, F,... Similarly, the spin angular momentum of each electron is quantised along the internuclear axis and gives rise to a total spin along the axis denoted by the symbol Σ. It is unfortunate that the symbol Σ is rather overworked, but the context in which it is used usually shows its meaning.

Just as in atoms we defined a total angular momentum $\mathbf{J} = \mathbf{L} + \mathbf{S}$, so we use a similar quantity, Ω, where

$$\Omega = \Lambda + \Sigma.$$

It is straightforward to extend the notation used for atomic states to those of diatomic molecules. Comparing that used for atoms, $^{2S+1}L_J$, with that for molecules, $^{2\Sigma+1}\Lambda_\Omega$, the meaning of the latter should be clear. For example, suppose we have $\Sigma = 1$ and

Fig. 5.12. The energy of a molecule can be partitioned into electronic (E), vibrational (V), and rotational (R) energy levels.

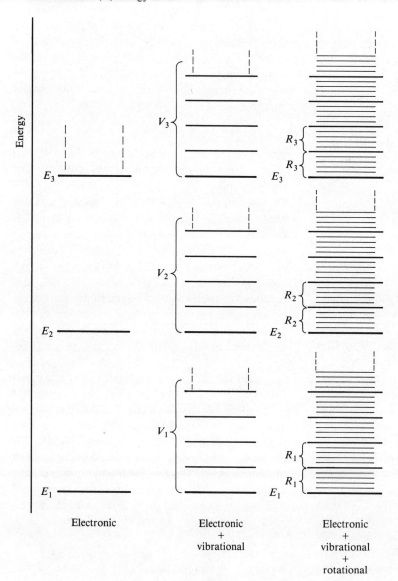

Table 5.2. *Principal selection rules for diatomic molecules*

$$\Delta \Lambda = 0, \pm 1$$
$$\Delta \Sigma = 0; \Delta S = 0$$
$$\Delta \Omega = 0, \pm 1$$
$$\Sigma^+ \leftrightarrow \Sigma^+, \Sigma^- \leftrightarrow \Sigma^-, g \leftrightarrow u \text{ allowed}$$
$$\Sigma^+ \leftrightarrow \Sigma^-, g \leftrightarrow g, u \leftrightarrow u \text{ forbidden}$$

There are further selection rules, which we ignore. These concern changes in angular momentum caused by motion of the nuclei.

$\Lambda = 2$, then Σ and Λ can be combined vectorially to give three values for Ω i.e. 1, 2, or 3. Thus we obtain the states $^3\Delta_1$, $^3\Delta_2$, $^3\Delta_3$. For homonuclear diatomics, g or u subscripts are added, these indicating the symmetry of the state with respect to inversion just as we used for describing σ and π orbitals. To complete the picture, for Σ states a right superscript is added as well. This consists of a + or a − sign showing whether the wavefunction belonging to the state is symmetric or antisymmetric with respect to reflection in a mirror plane drawn through the centres of the two nuclei.

We can use the hydrogen molecule to illustrate this. To keep things fairly simple we shall ignore the Ω values, and restrict ourselves to writing the term symbols. The ground state configuration is $1s\sigma_g^2$, which having electrons with spins paired, and each with $\lambda = 0$, so $\Lambda = 0$ gives rise to a $^1\Sigma_g^+$ term. The first excited state configuration, $1s\sigma_g\, 1s\sigma_u$, with two electrons again has $\Lambda = 0$, but $\Sigma = 0$ or 1, and gives $^1\Sigma_u^+$ or $^3\Sigma_u^+$. In similar fashion, the configuration $1s\sigma_g\, 2p\pi_u$ gives rise to $^1\Pi_u$ and $^3\Pi_u$ terms. Generally the triplet states are found to be of lower energy than singlets, but as shown by the selection rules in table 5.2, singlet to triplet transitions are forbidden. Therefore we should expect to find the first two lines in the spectrum of the hydrogen molecule to arise from the transitions shown in fig. 5.13.

If molecular spectroscopy were confined to the rather rarefied task of applying selection rules as an end in itself, it would not play such a prominent part in modern chemistry. The key to its popularity lies in its use in analytical work. We know from our examination of molecular orbitals that carbon–carbon double bonding gives rise to π bonding and

π^* antibonding orbitals. Transitions occur between these orbitals, written $\pi^* \leftarrow \pi$, at wavelengths of about 180 nm. This value is approximately constant for a single double bond, independent of the remainder of the molecule. Similarly a $\pi^* \leftarrow \pi$ transition occurs in carbonyl compounds, but they are also renowned for producing another transition which involves one of the non-bonding electrons on the oxygen atom being transferred to the π^* antibonding orbital. These transitions, labelled $\pi^* \leftarrow n$, are characteristic of carbonyl compounds and occur at around 280 nm. There is an interesting connection between the $\pi^* \leftarrow n$ transitions and the $d \leftrightarrow d$ transitions which give the colour to transition metal complexes. As we found in section 5.2, the $d \leftrightarrow d$ transitions are symmetry forbidden. So are $\pi^* \leftarrow n$ transitions. This is apparent if we realise that the non-bonding orbital on a carbonyl oxygen is effectively an oxygen 2p orbital. The π^* orbital is also built up from an oxygen 2p orbital. Our symmetry criteria for electric dipole transitions shows that the orbitals involved in the transition must have different parity. Hence $\pi^* \leftarrow n$ transitions are forbidden. The $d \leftrightarrow d$ transitions are similarly forbidden. Why then do they occur? The reason is that our classification of the orbitals in molecules has assumed the molecules to be of fixed geometry. In reality they vibrate and this has the effect of mixing in orbitals of different symmetries with one another. For example, d orbitals become contaminated with some p char-

Fig. 5.13. The origin of two lines in the spectrum of the hydrogen molecule. 'Forbidden' lines can occur if the spin selection rules do not hold good. Transitions to some excited states cause the molecule to break up; then bands instead of lines appear.

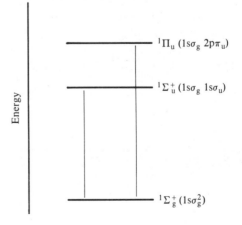

acter, and the n or π* orbitals take on some s character. The presence of orbitals of the opposite parity gives the all-important chance for the transitions to occur.

If we use electromagnetic radiation with sufficient energy it is possible to knock electrons out of orbitals involved in bonding and to put them into orbitals of very high energy which sometimes may be regarded as belonging to an individual atom rather than to the molecule as a whole. Such transitions are known as rydberg transitions. If even more energy is put in, electrons can be completely ejected from a molecule. This apparently perverse procedure is in fact extremely useful and forms the basis of photoelectron spectroscopy. A theorem which is due to Koopmans, and carries his name, says that the energy required to completely remove an electron from an orbital is equal to the energy of the orbital (but opposite in sign) (fig. 5.14). Not only does this allow us to measure the energies of orbitals to a reasonable accuracy it also means that we can use the method for analysis as in electron spectroscopy for chemical analysis (ESCA). This is because the energy of an orbital will depend on its electronic and nuclear environment. Fig. 5.15 illus-

trates this for carbon monoxide and carbon dioxide.

It is interesting to note that photoelectron spectroscopy makes use of the same type of practical technique that is used to study the photoelectric effect in metals. X-rays are used to bombard the molecules. Any ejected electrons are detected and

Fig. 5.15. ESCA spectra of carbon monoxide and carbon dioxide. The labels give the molecular orbitals involved in the transitions. Multiple lines are due to vibrations of the modules. Spectra taken from: K. Kimura *et al.*, *Handbook of HeI Photoelectron Spectra of Fundamental Organic Molecules*, Halsted Press, New York, 1981.

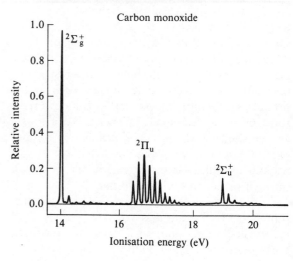

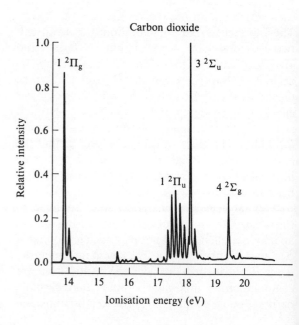

Fig. 5.14. The energy of a photon needed to completely remove an electron from orbital B is E_B. From Koopmans' theorem the energy of the orbital is $-E_B$.

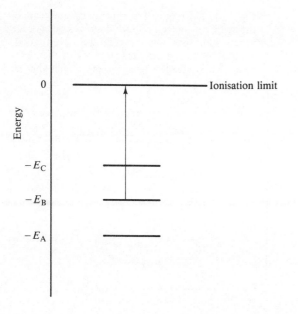

their energies measured. The number of electrons with a given energy are counted and plotted against the wavelength of the X-rays. Knowing the energy of the X-rays and the energy of the electrons, the difference gives the energy with which the electrons were bound to the atom in question, i.e. their orbital energy. The orbital energy plays an analogous part to the work function in the photoelectric effect.

Questions

5.9   (i) What is the ground state electron configuration of oxygen, O_2?

    (ii) Explain, as far as possible, how this configuration gives rise to the term $^3\Sigma_g^-$.

5.10 The first five electronic states of oxygen have the terms $^3\Sigma_g^-$, $^1\Delta_g$, $^1\Sigma_g^+$, $^3\Delta_u$, $^3\Sigma_u^+$ in order of increasing energy. Which transitions can occur between them?

5.11 Sketch the π and π^* orbitals for a carbonyl group. Also, show the non-bonding 2p orbital on the oxygen atom. Assume the $C{=}O$ bonds to be along the z-axis so that the π, π^* orbitals are built from $2p_x$ orbitals, and the non-bonding orbital is $2p_y$.

    Explain why the $\pi^* \leftarrow n$ transition is expected to be magnetic dipole allowed.

5.12 What would you predict to happen to the length, and strength, of the carbon–oxygen bond in a carbonyl group when it undergoes an $\pi^* \leftarrow n$ transition? Explain.

5.13 Write a wavefunction containing part s and part p character as $s + kp$. Write down the equation for the transition dipole moment, expand it to show the different parts and then use symmetry arguments to say which parts are non-zero.

5.4 / Vibrational spectroscopy

All molecules vibrate to a greater or lesser extent. If electromagnetic radiation of the correct energy engulfs a molecule then it may entice it to increase the energy with which it vibrates and the molecule will change from one vibrational energy level to another. As should be familiar now this can

Fig. 5.16. The four normal vibrations of carbon dioxide. The two bending modes are degenerate.

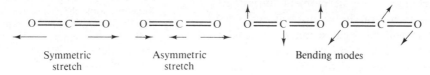

Symmetric   Asymmetric   Bending modes
stretch   stretch

Fig. 5.17. The symmetric stretching mode of carbon dioxide always has zero resultant dipole moment so this mode is not infrared active.

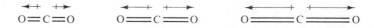

Fig. 5.18. During a bending mode of carbon dioxide there is a fluctuating dipole moment. Therefore this mode is infrared active.

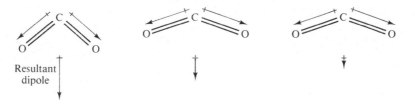

Resultant
dipole

only come about if there is a non-zero transition dipole moment. Although the motion of the atoms may appear extremely complicated, it happens that any combination of vibrational movements can be built up from a fixed number of fundamental vibrations. For a linear molecule of N atoms there are $3N$-5 fundamental vibrations; for a non-linear molecule there are $3N$-6. In general only some of the fundamental vibrations will give rise to dipole moments that can interact with the electromagnetic radiation. Those that do are said to be infrared active owing to the fact that transitions between vibrational energy levels occur in that part of the electromagnetic spectrum. Taking the carbon dioxide molecule as an example, its four fundamentals are shown in fig. 5.16. If we imagine the impossible situation (and it really is impossible, see section 6.11) where the molecule is not vibrating at all, it would have no net dipole moment owing to the two opposing dipole moments associated with the two carbon–oxygen bonds cancelling out (fig. 5.17). This is also the situation in the symmetric stretching mode, which as a consequence cannot be detected in vibrational spectroscopy. The asymmetric stretching mode is a different matter. If we draw the resultant dipole moment at various stages in the vibration, we obtain results like those in fig. 5.18. There is a change in dipole moment during the vibration so a transition dipole moment can occur and the mode will be infrared active. A similar statement is true of the asymmetric stretch.

If a molecule has very little vibrational energy its vibrations occur with simple harmonic motion (see M.5.2). The frequency of vibration is given by

$$f_0 = \frac{1}{2\pi} \sqrt{\left(\frac{k}{\mu}\right)}$$

where k is the force constant of the bond and μ is the reduced mass. The typical potential energy curve for simple harmonic motion is shown in fig. 5.19. Superimposed upon it are the energy levels which are the solutions of the Schrödinger equation for a simple harmonic oscillator. There are a number of intriguing things about the energy levels and wavefunctions.

First, the energy levels are given by

$$E_v = (v + \tfrac{1}{2})hf_0$$

where v is the vibrational quantum number.

Table 5.3. *Hermite polynomials*

The first five Hermite polynomials $H_v(s)$ are

$$H_0(s) = 1$$
$$H_1(s) = 2s$$
$$H_2(s) = 4s^2 - 2$$
$$H_3(s) = 8s^3 - 12s$$
$$H_4(s) = 16s^4 - 48s^2 + 12.$$

Further polynomials can be obtained by using the recursion relation

$$H_{v+1}(s) = 2sH_v(s) - 2vH_{v-1}(s).$$

The quantity s is given by

$$s = \sqrt{\left(\frac{4\pi^2 m f_0}{h}\right)}x$$

or, for convenience, $s = \sqrt{(\alpha)}x$.
The solution for the simple harmonic oscillator is

$$\psi_v(x) = N_v e^{-s^2/2} H_v(s)$$

where $N_v = \sqrt{\left(\dfrac{\sqrt{\alpha}}{2^v v! \sqrt{\pi}}\right)}$

with energies $E_v = (v + \tfrac{1}{2})hf_0$.

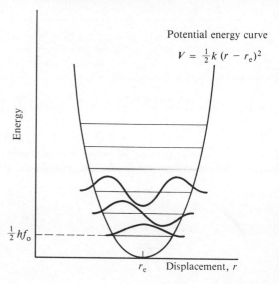

Fig. 5.19. The energy levels for a simple harmonic oscillator are evenly spaced. The zero point energy is $\tfrac{1}{2}hf_0$. Superimposed on the first three levels are their corresponding wavefunctions. Note that the wavefunctions are non-zero outside the classical potential energy curve. The equilibrium separation of the nuclei is represented by r_e. In one dimension the wavefunctions $\psi_v(x)$ are formed from Hermite polynomials. See table 5.3.

Potential energy curve

$$V = \tfrac{1}{2}k(r - r_e)^2$$

Energy

$\tfrac{1}{2}hf_0$

r_e Displacement, r

Spectroscopists often use wavenumbers as a measure of energy instead of joules. The unit of wavenumber is cm^{-1}. To convert to wavenumbers we take the energy in joules and divide by hc. In this case,

$$\varepsilon_v = (v + \tfrac{1}{2})\frac{hf_0}{hc} \quad (\text{cm}^{-1})$$

or

$$\varepsilon_v = (v + \tfrac{1}{2})\bar{\omega}_0 \quad (\text{cm}^{-1})$$

with $\bar{\omega}_0 = f_0/c$ being the frequency in wavenumbers. We shall use E_v or ε_v in future depending on which is the most useful in context.

Notice that when $v = 0$, the energy is $E_0 = \tfrac{1}{2}hf_0$. The fact that E_0 is not zero means that the energy of a simple harmonic oscillator can never be reduced below a minimum value, even at absolute zero. The quantity $\tfrac{1}{2}hf_0$ is the zero point energy. The appearance of the zero point energy is inexplicable in classical physics. Its explanation in quantum theory lies with Heisenberg's uncertainty principle. We shall pass this by for the present, but return to it in chapter 6.

Secondly, the difference in energy between any two adjacent vibrational levels is constant, so owing to the selection rule $\Delta v = \pm 1$, all transitions should occur at a single frequency. Therefore the spectrum should consist of a single line. For heteronuclear diatomic molecules this is broadly true, but in the majority of cases it is not. This is because the motion of atoms when vibrating is not strictly simple harmonic owing to the electrical interactions between the nuclei and electrons of the atoms at short and long bond lengths. If two atoms are squeezed very closely together, their electron clouds will repel very strongly. On the other hand, as the bond length increases to large values eventually the bond will break and the molecule will have dissociated into atoms. Fig. 5.20 compares the potential energy curves for a simple harmonic oscillator and a typical, real bond.

The minimum in the curves comes at the equilibrium separation, r_e. As we have already mentioned, for small changes about r_e the vibrations are simple harmonic. The mathematical form of a realistic potential energy curve is given to a good approximation by the Morse function

$$V_M = D_e[1 - e^{-a(r - r_e)}]^2.$$

D_e is the dissociation energy i.e. the energy needed to completely rupture the bond. The constant a is chosen to make V_M fit the observed potential energy as closely as possible. If the Schrödinger equation is solved with V_M as the potential energy term, the solutions are found to be

$$\varepsilon_v = (v + \tfrac{1}{2})\bar{\omega}_0 - (v + \tfrac{1}{2})^2 \chi\bar{\omega}_0 \quad (\text{cm}^{-1}).$$

The selection rules are also relaxed, being $\Delta v = \pm 1$, ± 2, ± 3,.... χ is the anharmonicity constant. Typically its value is of the order 10^{-2}. The larger the value of χ, the greater is the deviation from simple harmonic motion. As v becomes increasingly large the energy levels converge (fig. 5.21). By using the formulae for f_0 and ε_v, values of k, $\bar{\omega}_0$, and χ can

Fig. 5.20. The potential energy curve for a real molecule approximates to that for a simple harmonic oscillator only at small displacements about the equilibrium displacement, r_e.

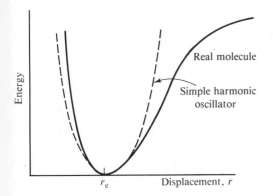

Fig. 5.21. For an anharmonic oscillator the energy levels are unevenly spaced and gradually converge.

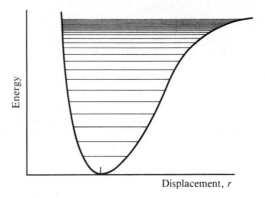

be calculated from the observed frequencies of the lines in the vibrational spectrum. This task forms the core of question 5.16.

If we return to looking at fig. 5.19 we see that the wavefunctions for a simple harmonic oscillator are finite beyond the boundaries of the potential energy curve. Classically the curve sets the absolute boundary to the motion of a simple harmonic oscillator. However, where the wavefunction is non-zero, so is its square. Therefore quantum theory says that there is a finite probability of finding a mass executing simple harmonic motion outside the classically allowed region. Classically it has insufficient energy to appear beyond the limits set by the potential energy curve. In quantum theory the mass is said to tunnel through the barrier. Tunnelling is believed to occur in a variety of circumstances, some

Fig. 5.22. Vibrations of the atoms in ammonia can bring about the inversion of the molecule.

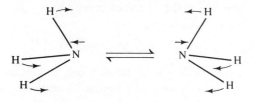

Fig. 5.23. The potential energy curves for the inverted and non-inverted forms of ammonia intersect.

of which are the subject of the questions that follow; but a particularly interesting example arises in the vibrations of ammonia.

The vibrational spectrum of ammonia shows a series of lines which occur in pairs and are associated with the vibrations shown in fig. 5.22 which bring about the inversion of the molecule. The inversion occurs very easily, at a frequency of about 2000 MHz. If the inversion were not to occur we should expect there to be one potential energy curve, approximately that of a simple harmonic oscillator. With inversion there are two such curves (fig. 5.23), but they intersect at a point corresponding to the configuration in which the nitrogen and three hydrogens all lie in the same plane. If we concentrate on one of the lowest vibrational levels of the first potential well and superimpose its wavefunction as in fig. 5.24, we find that the wave-function tunnels through the potential barrier and can enter the region of the second potential well. Similarly a wavefunction belonging to the second well can penetrate to the first. We can interpret this by saying that although the molecule may be vibrating in the non-inverted form, there is a finite probability of its

Fig. 5.24. A wavefunction belonging to one of the double potential wells can tunnel through to the other. Interference can take place between wavefunctions belonging to the individual wells thus giving rise to a doubling of the energy levels. The diagram is not drawn to scale.

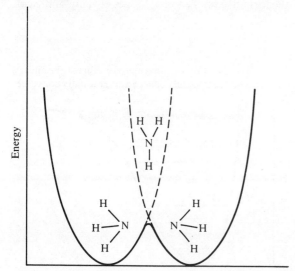

Displacement of the hydrogen atoms from the nitrogen atom

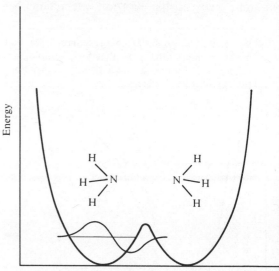

Displacement of the hydrogen atoms from the nitrogen atom

suffering an inversion and finding itself vibrating in the inverted form. Similarly, the reverse process can occur.

We shall label the wavefunctions belonging to the first and second well as ψ_1 and ψ_2 respectively. As each can penetrate into the other's well, we would expect them to show either constructive or destructive interference to give the resultants $\psi_+ = \psi_1 + \psi_2$ and $\psi_- = \psi_1 - \psi_2$. The energies of ψ_+ and ψ_- will be different, thus giving rise to a pair of energy levels E^+, E^-. It is found that E^+ lies lower than E^-. Thus each vibrational energy level for the single wells gives rise to a pair of energy levels with transitions

Fig. 5.25. Inversion doubling in ammonia gives rise to pairs of energy levels shown as E^+ and E^-. The numbers in brackets are the values of the vibrational quantum numbers.

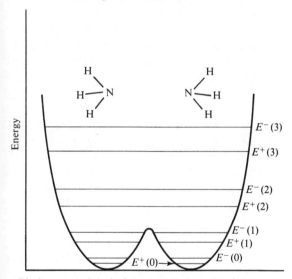

Displacement of the hydrogen atoms from the nitrogen atom

Fig. 5.26. If a spherical distribution of charge is disturbed, then a dipole moment can be induced. The easier it is to disturb the distribution of charge, the higher is the polarisability.

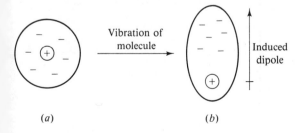

(a)   *(b)*

allowed between ψ_+ and ψ_- states. This explains the occurrence of pairs of lines in the spectrum, a phenomenon which for obvious reasons is called inversion doubling (fig. 5.25).

Before leaving vibrational spectroscopy we shall take a brief diversion to examine Raman spectroscopy. Raman spectroscopy allows us to obtain information about vibrations which do not show up in 'ordinary' vibrational spectroscopy. It does so by making use of an induced dipole moment with which the electromagnetic radiation can interact. Suppose there is a totally symmetric arrangement of the electrons and nuclei in a molecule, which we can depict in an idealised way by a sphere of charge. Fig. 5.26(*a*) shows a cross section through the sphere. Now suppose that the molecule undergoes a vibration which causes a disturbance to the spherical distribution, as shown in fig. 5.26(*b*). There is a dipole moment set up, which has been induced by the manner of the vibration. The magnitude of the induced dipole is given from

$$\mu = \alpha F$$

where F is the electric field of the incoming radiation and α is the polarisability. Strictly α should be treated as a tensor but we need only concentrate on the point that the induced dipole can interact with the electromagnetic radiation and cause transitions to occur. Thus molecules such as the homonuclear diatomics which do not have a pure vibrational spectrum will give lines in a Raman spectrum. Incidentally, a Raman experiment involves the observation of light scattered from molecules. The scattered light is detected at right angles to the direction of the incoming electromagnetic field. Generally, the intensity of the scattered radiation requires lasers to be used as the source of light if the spectrum is to be observed.

Questions

5.14 Which of the following molecules can have an infrared spectrum?

$NH_3, HCl, H_2, CO_2, O_2, Cl_2, C_2H_4.$

Explain your answer.
    Which would you expect to give rise to a Raman spectrum?

5.15 Using the formula for the energy of an anharmonic oscillator, work out the difference in energy, ΔE, between two adjacent levels with vibrational quantum numbers v and $v+1$. Hence show that, if v is large enough, $\Delta E = 0$, thus proving that the energy levels gradually become increasingly close together.

5.16 The transition $v = 1 \leftarrow v = 0$ gives rise to the fundamental line in a vibrational spectrum. The transition $v = 2 \leftarrow v = 0$ gives the first overtone. For carbon monoxide the fundamental occurs at approximately $2143 \, \mathrm{cm}^{-1}$ and the first overtone at $4260 \, \mathrm{cm}^{-1}$.

Use the formula for the energy of an anharmonic oscillator to calculate (i) the anharmonicity constant, χ, (ii) $\bar{\omega}_0$.

Now use your result for $\bar{\omega}_0$ to calculate the force constant for carbon monoxide. First convert from wavenumber to frequency (Hz) by multiplying by the speed of light in $\mathrm{cm \, s}^{-1}$. Use the formula for f_0, p. 92, and note that the mass should be the reduced mass, $\mu = (m_1 m_2)/(m_1 + m_2)$. The mass of $^{12}_{6}\mathrm{C}$ is $1.993 \times 10^{-26} \, \mathrm{kg}$ and of $^{16}_{8}\mathrm{O}$ is $2.656 \times 10^{-26} \, \mathrm{kg}$.

5.17 (i) Is the reduced mass of $^{1}_{1}\mathrm{H}^{35}_{19}\mathrm{Cl}$ (hydrogen chloride) greater or less than that of $^{2}_{1}\mathrm{H}^{35}_{17}\mathrm{Cl}$ (deuterium chloride)?
   (ii) Similarly, how will their fundamental frequencies of vibration compare?
   (iii) Sketch the arrangement of the first few vibrational energy levels of these two molecules on the same potential energy diagram. How will their dissociation energies compare?
   (iv) In general, what effect will the presence of isotopes have on vibrational spectra?

5.18 The potential barrier preventing the escape of α-particles from the nuclei of atoms has the shape shown in fig. 5.27. According to classical physics an α-particle should have an energy at least equal to E_b if it is to get over the energy barrier and reach the outside world. How is it that experiment shows that in practice, α-particles are found to emerge from atoms with less energy than E_b?

5.19 Just as an electric dipole transition occurs only if the integral $\int \psi_2 \hat{\mu} \psi_1 \, dv \neq 0$, so induced dipole transitions only occur if $\int \psi_2 \hat{\alpha} \psi_1 \, dv \neq 0$. Here $\hat{\alpha}$ is the polarisability operator. For our present purposes we can ignore its precise form and concentrate on its symmetry properties. The main one is that it behaves like a product of two Cartesian coordinates or unit vectors. The latter are of odd parity so the polarisability must have odd $\times$ odd = even parity.

What are the selection rules for the symmetries of ψ_1 and ψ_2?

In a molecule with a centre of symmetry, which has 'g' and 'u' type wavefunctions, will all vibrational modes be infrared and Raman active at the same time?

5.20 The splitting between the two lowest energy levels for the ammonia inversion, $E^+(0)$ and $E^-(0)$, is approximately $0.8 \, \mathrm{cm}^{-1}$. What is the frequency of the transition (in Hz)? This gives a measure of the frequency of the inversion. How many times a second does the inversion occur?

5.21 Show that an electric dipole transition is allowed for a transition between the two

Fig. 5.27. Classically, if an α-particle is to leave the nucleus of an atom it must have an energy at least equal to E_b if it is to get over the energy barrier.

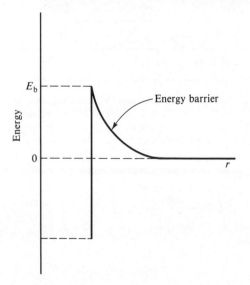

vibrational states with $v = 0$, and $v = 1$ of a simple harmonic oscillator.

Similarly, show that a transition between states with $v = 0$ and $v = 2$ is forbidden. Use the standard integrals in appendix A to help you. Assume that the dipole moment operator is $\hat{\mu} = -ex$.

5.22 In Fig. 5.28(a) are shown the potential energy curves for the electronic ground state and an excited state of a molecule. In the lower state two vibrational levels, $v'' = 0$, $v'' = 4$, are shown with their corresponding probability density distributions.

(i) For $v'' = 0$, what is the most probable internuclear distance?

(ii) For $v'' = 4$, what are the most probable internuclear distances?

Show your answers by marking the diagram accordingly.

A very important principle, the Franck–Condon principle, says that electronic transitions occur so quickly that a molecule does not have time to change its geometry while the transition occurs. This means that the internuclear separation is constant during an electronic transition. Sometimes this is known as a vertical transition.

Apply the Franck–Condon principle to answer the remaining parts of the question.

(iii) If the molecule with $v'' = 0$ undergoes an electronic transition to the upper state, which vibrational level is it most likely to find itself in?

(iv) Suppose the molecule originally has $v'' = 4$, what might happen?

(v) Now look at fig. 5.28(b) and answer (iii) and (iv) in this case.

5.23 Think back to how we dealt with the solutions of Schrödinger's equation for the one- and two- and three-dimensional box problems. Write down expressions for the energies and wavefunctions for a two-dimensional simple harmonic oscillator. See table 5.2 for the one-dimensional solutions.

Discuss the possibility of degeneracy for

solutions having energies hf_0, $2hf_0$, $3hf_0$, and $4hf_0$.

5.5 Rotational spectroscopy

If a molecule is to be excited by electromagnetic radiation to rotate with greater energy then it

Fig. 5.28. The electronic and vibrational levels for a ground state having two different excited states. Figures (a) and (b) are referred to in question 5.22.

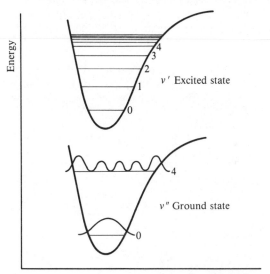

(a)

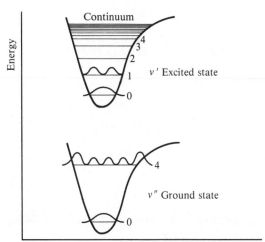

(b)

must have a dipole moment. As we are considering the rotation of molecules as a whole we are not concerned with dipole moments brought about by vibrations. Rather, to show a pure rotation spectrum a molecule must have a permanent dipole moment. For example, HCl, CO, NH_3, H_2O will have a pure rotation spectrum while H_2, O_2, N_2, CH_4 will not.

If we ignore their vibrations the heteronuclear diatomic molecules serve as examples of plane rigid rotators. Mathematical details of plane rigid rotators are to be found in M.5.3. Here we shall

Fig. 5.29. The rotational motion of a particle in the xy-plane is analysed in terms of the azimuthal angle, ϕ. The coordinates of the particle are $x = r\cos\phi$; $y = r\sin\phi$.

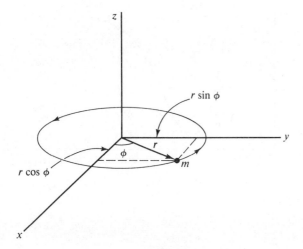

concentrate on the result which quantum theory gives for the energy levels. This is

$$E_J = \frac{h^2}{2I} J(J+1)$$

or

$$E_J = BJ(J+1)$$

where $B = h^2/2I$ is the rotational constant. J is the rotational quantum number whose values are limited to the integers $0, 1, 2, \ldots$, I is the moment of the inertia. The selection rule is $\Delta J = \pm 1$, and using this it is easy to show that the lines in the spectrum should be evenly spaced $2B$ apart.

At the beginning of this chapter we noted that the energy difference between rotational levels is much less than that between vibrational levels. Owing to the small differences between rotational levels it happens that at room temperature molecules will not be confined to the lowest, or ground state, rotational level (see M.5.4). For this reason the lines in a rotational spectrum have markedly different intensities. If, for example, the state with $J = 3$ is more highly populated than that with $J = 1$, transitions from the $J = 3$ to another state are inherently more likely than those from the $J = 1$ state. This effect can be seen clearly in the rotational spectrum of carbon monoxide (fig. 5.30).

Not surprisingly other complications also occur. When a molecule rotates with higher and higher energy, the atoms are thrown out away from one another. This causes the bonds to lengthen and the moment of inertia to increase. Also, real mole-

Fig. 5.30. The rotational spectrum of carbon monoxide (spectrum taken by the author).

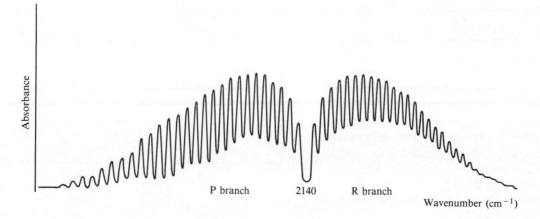

cules do vibrate and the constantly changing bond length also affects the moment of inertia. In these circumstances the expression for E_J becomes

$$E_J = BJ(J+1) - DJ^2(J+1)^2$$

where D is a small, positive constant called the centrifugal distortion constant. Just as vibrations

Fig. 5.31. For each vibrational level there is a corresponding set of much more closely spaced rotational levels.

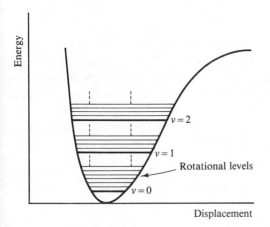

Fig. 5.32. The diagram shows typical transitions between rotational states during a transition between two vibrational levels. It is convention that upper states are labelled with one prime v', J', while lower states have double primes, v'', J''.

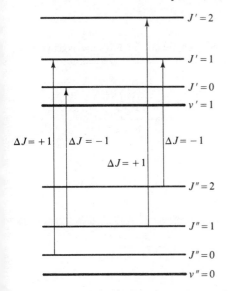

have an influence on rotational spectra, so rotations influence vibrational spectra.

Each vibrational state has its own set of rotational states, a few of which are shown in fig. 5.31. If we look at the energy levels more closely it becomes clear that the types of transitions are allowed by the selection rules $\Delta v = \pm 1$ and $\Delta J = \pm 1$ with vibrational and rotational transitions able to occur together (fig. 5.32). This gives rise to spectra like that for carbon monoxide shown in fig. 5.30. The lines for which $\Delta J = -1$ form the P branch of the spectrum, those with $\Delta J = 1$ form the R branch. In rare cases a diatomic molecule can also have a Q branch for which $\Delta J = 0$. A molecule which does possess a Q branch is nitrogen monoxide. It occurs because nitrogen monoxide has a component of orbital angular moment along the bond axis. Although it is unusual in diatomics, Q branches are found quite often in the spectra of polyatomic molecules. Indeed, the spectra of such molecules can be complicated owing to their having different moments of inertia about different axes of symmetry. These complications are dealt with in most books on microwave spectroscopy; we shall not discuss them here.

Questions

5.24 What is the physical significance of the fact that the energy levels of a plane rigid rotator are degenerate?

5.25 In M.5.3 we used $\psi(\phi) = A \cos(\phi)$ as a solution of the Schrödinger equation of a plane rigid rotator. If $\psi(\phi)$ is to act as probability density function, we must have $\int_0^{2\pi} |\psi(\phi)|^2 \, d\phi = 1$. Use this to determine the value of A, which is the normalisation constant.

5.26 Assume that $^1_1H^{35}_{17}Cl$ and $^2_1H^{35}_{17}Cl$ are plane rigid rotators. How do their rotational energy levels compare?

5.27 Write down the relation for the energy of (i) the vibrational levels, (ii) the rotational levels of a diatomic molecule (not a plane rigid rotator).

Now write down the relation for the total vibrational and rotational energy.

Use this to give a general formula for the

energy of a vibration–rotation transition. Assume $\Delta v = +1$, $\Delta J = +1$.

5.28 Neglect degeneracy and use the Boltzmann distribution (M.5.4) to estimate the difference in populations between
  (i) two electronic states,
  (ii) two vibrational states,
  (iii) two rotational states,
  (iv) two nuclear spin states.
Use the order of magnitudes for the energy separations involved in table 5.1. Assume the temperature to be 298 K (25 °C).

5.6 Electron spin resonance spectroscopy

Electrons possess the property which we have called spin and have a magnetic moment as a consequence. If we think classically for a moment, we could imagine an electron to be spinning in a clockwise or anticlockwise sense. We can associate this notion with the two values $m_s = \frac{1}{2}, -\frac{1}{2}$ for an electron corresponding to α and β spins. As a rule, states with $m_s = +\frac{1}{2}$ are degenerate with states having $m_s = -\frac{1}{2}$, but in the presence of a magnetic field the degeneracy is lifted. A state with $m_s = +\frac{1}{2}$ has a different energy from one with $m_s = -\frac{1}{2}$.

Fig. 5.33. For $s = \frac{1}{2}$ there are two cones of orientations, one having a projection of s_z or $m_s = +\frac{1}{2}$, and one with projection s_z or $m_s = -\frac{1}{2}$.

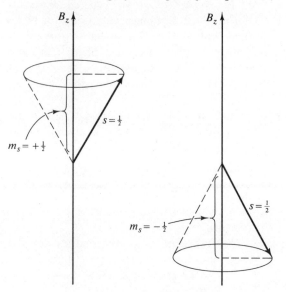

Classically we can imagine this as a reflection of the tendency of a magnet to align itself in a magnetic field. If a magnet lines up with the direction of its magnetic moment in the same direction as the field, it will have a lower energy than a magnet with its magnetic moment pointing against the direction of the field.

Electrons are not classical magnets. Each electron has a spin quantum number, $s = \frac{1}{2}$, which gives a total spin angular momentum $\sqrt{[s(s+1)]}\hbar$ or $\sqrt{(3/2)}\hbar$. The values of m_s give the projections of s on a given axis. The direction of an applied magnetic field destroys the isotropic nature of space and serves to define such an axis. Convention has it that the field defines the z-direction. Although m_s gives the projections of s on the z-azis, we cannot be sure of the precise orientation of the vector belonging to s. It forms a cone of orientations and precesses about the z-axis at the Larmor frequency (fig. 5.33).

The energy of an unpaired electron, in a magnetic field, **B**, is given by

$$E = -\mathbf{m}\cdot\mathbf{B}.$$

m is the magnetic moment vector. It is related to the spin vector **s** by

$$\mathbf{m} = -\frac{2\pi}{h}g_e\mu_B\mathbf{s}.$$

g_e is the electron g-value and $\mu_B = eh/2m_e$ is the Bohr magneton.

Fig. 5.34. In a magnetic field, B_z, the degeneracy of electron spin states α and β is lifted.

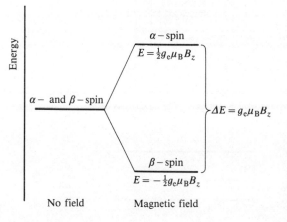

When the magnetic field is in the z-direction,

$$E = -m_z B_z$$

$$= \frac{1}{\hbar} g_e \mu_B s_z B_z.$$

But s_z we can recognise as the projections on the z-axis which we know to be $+\frac{1}{2}\hbar$ and $-\frac{1}{2}\hbar$. Thus, the two possible energies (fig. 5.34) are

$$E_1 = \tfrac{1}{2}g_e \mu_B B_z \qquad \text{(for } \alpha\text{-spin)}$$

and

$$E_2 = -\tfrac{1}{2}g_e \mu_B B_z \qquad \text{(for } \beta\text{-spin).}$$

The difference in energy is

$$\Delta E = g_e \mu_B B_z.$$

If we apply Planck's equation it should be possible to find a frequency, f, for which

$$hf = g_e \mu_B B_z$$

or

$$f = \frac{1}{h} g_e \mu_B B_z.$$

Radiation of this frequency should bring about transitions between the state with $m_s = -\frac{1}{2}$ to that with $m_s = +\frac{1}{2}$. That is, cause the spin to change from β to α.

This is the basis of electron spin resonance (e.s.r.) spectroscopy. The wavelength of radiation which brings about a transition of an electron from one spin state to another occurs at a wavelength of about 3 cm at magnetic field strengths of a few tenths of a tesla. For transitions to occur the frequency of the radiation has to be equal to the Larmor frequency of the precession of the spin vector about the z-axis. Then the transfer of energy from the radiation to the electron is the most efficient. Then a resonance condition is met, hence the name of this type of spectroscopy.

The magnitude of the absorption frequency depends on the strength of the magnetic field. In practice it is easier to keep the frequency of the microwave radiation constant and vary the field strength until resonance, and a strong absorption of the radiation, occurs. Instead of the usual hump-shaped absorption curve, the output from an e.s.r. spectrometer is given as the first derivative of the ordinary absorption curve (fig. 5.35).

For a free electron $g_e = 2.0023$, but for an unpaired electron in a free radical or ion its g-value will differ from this value. An electron's g-value depends on the environment in which it finds itself. To understand why we must not forget that a magnetic field acting on a system of charges, such as electrons, will cause them to move in a circular path. This will provide the electrons with some angular momentum even if they had none in the absence of the field. The induced motion also produces a magnetic field which, in accordance with Lenz's law, opposes the applied field. The electron undergoing an e.s.r. transition is therefore subject to two fields: the applied (external) field and the induced (local) field. The net field the electron experiences will, in general, be less than the applied field. However, this is not always the case for, especially in transition metal complexes, the local field can augment the applied field.

If we think of the g-value as a factor multiplying the field we can re-arrange our equation for the frequency as

$$f = \frac{1}{h}\mu_B(g_e B_z)$$

for a free electron. In a molecular environment the factor in brackets is usually, but not always, less than $g_e B_z$. Accordingly we put

$$f = \frac{1}{h}\mu_B(g B_z)$$

Fig. 5.35. The e.s.r. signal is the first derivative of an ordinary absorption curve.

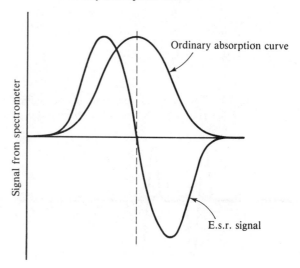

Ordinary absorption curve

Signal from spectrometer

E.s.r. signal

Table 5.4. *The spins of some nuclei*

Atom	Spin	Atom	Spin
$^{1}_{1}H$	$\frac{1}{2}$	$^{14}_{7}N$	1
$^{2}_{1}H$	1	$^{15}_{7}N$	$\frac{1}{2}$
$^{13}_{6}C$	$\frac{1}{2}$	$^{16}_{8}O$	0
$^{12}_{6}C$	0	$^{19}_{9}F$	$\frac{1}{2}$

where the difference between g and g_e reflects the impact of the local field.

In addition to the variation of the g-value from one species to another, there is another very important feature of e.s.r. spectra. It is nearly always the case that the lines are split into two or more further lines. This can happen when there are two or more unpaired electrons present, and we say that the spectrum shows fine structure. However it is uncommon for molecules to have several unpaired electrons. Far more often the spectra display hyperfine structure. This occurs owing to the nuclei of atoms consisting of collections of neutrons and protons each with their own magnetic moments. A neutron and a proton each have a nuclear spin

quantum number, I, of $\frac{1}{2}$ which gives rise to two spin states $m_I = +\frac{1}{2}$ and $m_I = -\frac{1}{2}$. (Like electrons, they are fermions; see question 3.17.) The individual spins of the neutrons and protons sum to give a total spin which varies from one nucleus to another (table 5.4). For a spin I there will be $2I + 1$ values of m_I. A nuclear magnetic field will envelop an electron that spends time in the vicinity of the nucleus. Thus, again, the field which the electron feels is not owing only to the applied field, nor to the induced field.

If we take the methyl radical, $CH_3\cdot$, as an example, the carbon atom has zero nuclear spin while the hydrogen atoms, i.e. the protons, have a spin of $\frac{1}{2}$ each. In a given collection of methyl radicals any one radical may have its protons in a state with $m_s = +\frac{1}{2}$ or $m_s = -\frac{1}{2}$. As for electrons, we shall label these as α- and β-spin states. There are eight possible combinations of α- and β-spins for the three protons as shown in fig. 5.36.

The effective spins are 3α, α, β, 3β in the ratio $1:3:3:1$. These four different combinations produce four different magnetic fields of which two will enhance the applied field and two reduce it. The spectrum of $CH_3\cdot$ does indeed consist of four lines with the expected intensities (fig. 5.37).

The protons in $CH_3\cdot$ are all in a similar environment because the radical is planar. The protons are said to be equivalent. In general the influence of n equivalent protons is given by the coefficients in the binomial expansion of $(1 + x)^n$. The coefficient of x^n is $n!/[r!(n - r)!]$, but these values

Fig. 5.36. The eight combinations of spins for three equivalent protons give 3α, α, β and 3β in the ratio $1:3:3:1$.

Proton	Spin combinations							
1	α	α	α	β	β	β	α	β
2	α	α	β	α	α	β	β	β
3	α	β	α	α	β	α	β	β
Total	3α	α	α	α	β	β	β	3β

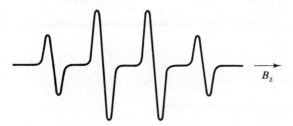

Fig. 5.37. The idealised spectrum of the methyl radical, $CH_3\cdot$, consists of four lines in ratio of intensities $1:3:3:1$.

Fig. 5.38. The members of Pascal's triangle give the splitting pattern for a system of equivalent protons. The numbers of αs and βs is given by the integers n, $n - 2$, $n - 4, \ldots, 0$, but zero must be counted once only. For example, for four protons we have

$$4\alpha : 2\alpha : 0 : 2\beta : 4\beta :: 1 : 4 : 6 : 4 : 1$$

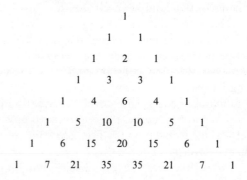

are more accessible in the form of Pascal's triangle displayed in fig. 5.38.

The magnitude of the gap between the hyperfine lines gives the magnitude of the hyperfine coupling constant, which is a characteristic of the nucleus causing the splittings. The appearance of an e.s.r. spectrum, taken together with the evaluation of the hyperfine coupling constants, can be used as a means of characterising the free radical or ion investigated. This is a theme of some of the questions that follow.

Questions

5.29 Which of the following would not give an e.s.r. spectrum? Briefly explain

$$CH_3^+, \quad H_2, \quad H\cdot, \quad H^-, \quad NO_2, \quad SO_3^-,$$
$$CH_3\dot{C}HOH.$$

Of the others, which would give just one line in the spectrum?

5.30 What would be the appearance of the spectrum of $CD_3\cdot$? $(D \equiv {}_1^2H)$.

5.31 Predict the appearance of the spectrum of the benzene anion, $C_6H_6^-$. Is there a reason why you would find it surprising that it should show hyperfine structure? (Hint: what is the nature of the orbital into which the extra electron goes?)

5.32 What would you predict for the spectrum of the ethyl radical, $C_2H_5\cdot$? Note that this radical consists of a CH_3 and a CH_2 group. Thus there is one set of three equivalent protons and another set of two equivalent protons.

5.33 Naphthalene, $C_{10}H_8$, has ten π electrons which fit into the first five molecular orbitals derived by Hückel theory. The extra electron which forms the anion $C_{10}H_8^-$ goes into the molecular orbital $\psi_6 = 0.0425(10\phi_1 + 10\phi_4 + 6\phi_6 + 6\phi_7 - 6\phi_2 - 6\phi_3 - 10\phi_5 - 10\phi_8)$. The coefficients are only approximate. See fig. 5.39 for the number system.
  (i) Which carbon atoms have equal charge densities?
  (ii) Which sets of carbon atoms are equivalent to one another?

  (iii) Which protons are equivalent in the influence they will have on the unpaired electron?
  (iv) What will be the pattern of splittings in $C_{10}H_8^-$?

5.34 Suppose a free electron is subject to microwave radiation of frequency 9000 MHz. At what field strength would (i) a free electron, (ii) an electron with g-value 2.000, suffer a transition?

5.35 The Boltzmann distribution gives the relative populations of two energy levels separated by an energy ΔE as $n_1 = n_0 e^{-\Delta E/kT}$. (Also see M.5.4.) What is the ratio of the populations of spin state α to spin state β for a free electron in a field of 0.3 T?

Does this difference in populations mean that e.s.r. is of low, moderate or high sensitivity?

5.36 Spin–orbit coupling has an impact on e.s.r. The spin–orbit coupling energy is given by

$$E_{s.o} = \frac{1}{\hbar^2}\zeta \mathbf{s}\cdot\mathbf{l}$$

where $\mathbf{l}$ is the orbital and $\mathbf{s}$ the spin angular momentum vector. ζ is the spin–orbit coupling constant. (Also, see section 5.2.)
  (i) Which gives the lower energy, $\mathbf{s}$ and $\mathbf{l}$ parallel or anti-parallel?
      It so happens that $\mathbf{l}$ acts in the opposite direction to the applied field.
  (ii) With l_z acting against B_z, does the upper, α, spin state of fig. 5.34 suffer a rise or fall in energy?
  (iii) Likewise, what happens to the lower, β, state?

Fig. 5.39. The numbering system for naphthalene.

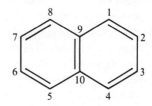

(iv) Redraw fig. 5.34 showing the pattern of energy levels with spin–orbit coupling at work. Will the resonant frequency be raised or lowered when spin–orbit coupling acts?

5.7 Nuclear magnetic resonance

In a magnetic field a nucleus with a spin of $\frac{1}{2}$ behaves very much like an electron. Its α- and β-spin states will split apart in energy. Given the correct combination of magnetic field strength and frequency of electromagnetic radiation, transitions between spin states can be induced. This is the basis of nuclear magnetic resonance, n.m.r.

The simplest case is that of a bare proton. Owing to its having an opposite charge to that of an electron its magnetic moment is in the opposite sense to an electron's. Thus in the presence of a magnetic field the α-spin state lines are lower in energy than the β state. Transitions occur at a frequency

$$f = \frac{1}{h} g_N \mu_N B_z.$$

g_N is the nuclear g-value and $\mu_N = eh/2m_p$ is the nuclear magneton with m_p the proton mass. A quick calculation shows that at the same field strength a transition between proton α- and β-spin states occurs at a frequency some 800 times lower than in a free electron e.s.r. transition. This places n.m.r. as a radiofrequency experiment. As in e.s.r. the experimental method is to keep the frequency constant and vary the field strength until resonance occurs. Commonly frequencies of 60 MHz up to several hundred MHz are used, requiring field strengths between 1T and 10 T. In n.m.r. the spectrum of a molecule consists of a series of absorption curves, usually displaying fine structure in their splitting patterns.

Protons in a molecule are subject to magnetic fields arising from the circulating electrons and neighbouring nuclei. Both effects we found in e.s.r. as well. However, the first is dealt with in a different way to that in e.s.r. We know that the magnetic field experienced by a proton in a molecule will not be the same as the applied field. We write the effective field it experiences as B_{eff}. The proton will be screened by an amount $B_z - B_{eff}$. The extent of the screening is proportional to B_z and can be put equal to σB_z where σ is the screening constant. Then

$$\sigma B_z = B_z - B_{eff}$$

σB_z is the chemical shift. If the screening of a given proton is slight then $B_{eff} \approx B_z$ and, as we expect, the screening constant will be small. Similarly the chemical shift is small.

If we were to look at a proton attached to another atom as in a C—H or O—H bond, the proton in the O—H bond will be less well shielded than the proton in the C—H bond. This is because the oxygen with its greater electronegativity withdraws electron density from around the proton more efficiently than does a carbon atom. The chemical shifts of the two protons will be different and each will resonate at a different field strength.

Unfortunately, the definition of chemical shift as σB_z means that its value changes depending on the field strength used. For example, the same proton would have a different chemical shift in a 60 MHz experiment and a 300 MHz experiment. It is better to use values which are independent of the field. There are two common scales that do this. Both measure chemical shifts with respect to a given standard: the resonance of the protons in tetramethylsilane (TMS), $Si(CH_3)_4$. If the TMS resonance occurs at a field B_{TMS}, then the δ-scale is defined by

$$\delta = \frac{B_z - B_{TMS}}{B_{TMS}} \times 10^6.$$

This scale is dimensionless, and the factor 10^6 is included to bring most values of δ within the range 0–10. The τ-scale is used by those who have an aversion to negative numbers. The τ-scale changes the origin so that the TMS resonance is given a value of 10 and $\tau = 10 - \delta$; τ-values for protons in some common environments are collected in table 5.5.

In order to interpret n.m.r. spectra we must first examine the influence of neighbouring nuclei on the field a proton experiences. You should have little difficulty in recognising why the n.m.r. spectrum of pure methanol has the appearance shown in fig. 5.40. Generally n equivalent protons will split an absorption on a neighbouring atom into $n + 1$ lines with relative intensities given from Pascal's triangle. The quartet of lines in the methanol spectrum is the

Table 5.5. *Typical τ-values for protons in different environments*

Proton type	τ
R—CHO	0.5
C_6H_6	3.5
R—OH	4.0
$>C=CH_2$	5.5
—C≡CH	7.5
—CH_3	8.5

There is a wide variation from these values in some cases.

hydroxyl proton split by the three equivalent methyl protons. The doublet is the result of the methyl protons split by the hydroxyl proton.

The appearance of n.m.r. spectra is often more complex than those that we shall meet here. The reason for this is that the local field is not just dependent on the neighbouring nuclei. The field will depend on the total magnetic environment, which includes the electron spin density at the nucleus. Terms in the Hamiltonian which take account of these important but subtle hyperfine interactions

Fig. 5.40. The idealised n.m.r. spectrum of methanol. The ratio of areas under the peaks is 1:3 indicating the ratio of the two types of protons in methanol.

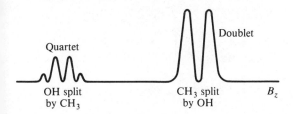

Fig. 5.41. The idealised n.m.r. spectrum of the compound $C_6H_{14}O$ of question 5.38. The signals are not to scale.

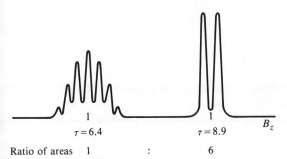

are difficult to derive. Further information can be obtained by consulting the books in the bibliography.

Some of the questions that follow illustrate the power of n.m.r. as a technique of chemical analysis.

Questions

5.37 Predict the appearance of the n.m.r. spectrum of pure ethanol.

5.38 An idealised impression of the n.m.r. spectrum of a compound $C_6H_{14}O$ is shown in fig. 5.41. The relative areas under the peaks give a measure of the numbers of protons of each type. Try to identify the compound.

5.39 We know that many nuclei have spins. See table 5.4. Of these $^{13}_6C$, $^{19}_9F$ and $^{31}_{15}P$ are sometimes used in addition to ^1_1H.
   (i) Estimate the abundance of $^{13}_6C$ compared to $^{12}_6C$? Why is it that $^{13}_6C$ n.m.r. had to await significant advances in electronics before it became widely used?
   (ii) What would be the effect on the $^{13}_6C$ resonance in $CHCl_3$ and $CDCl_3$ $(D \equiv {}^2_1H)$?

5.40 Temperature has a marked effect on some n.m.r. spectra. That of cyclohexane is a case in point. Fig. 5.42 shows the spectrum of cyclohexane at two temperatures. Use an organic textbook to study the various geometries of cyclohexane. Explain why, at lower tempera-

Fig. 5.42. The n.m.r. spectrum of cyclohexane shows two peaks at 'low' temperatures which merge into a single peak at about 330 K.

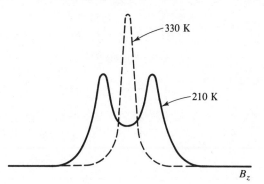

tures, there appear to be two sets of equivalent protons, while at higher temperatures all the protons appear to be equivalent.

5.41 A photon which induces a change of spin from β to α (or vice versa) means that s_z changes by $-\frac{1}{2}\hbar$ to $\frac{1}{2}\hbar$.
   (i) What is the net change in angular momentum?
   (ii) By the law of conservation of angular momentum, what must be the spin of a photon?
   (iii) What is the name given to species with this value of spin?

5.8 Summary

We have seen that quantum theory can ac- count for the spectroscopic properties of atoms and molecules. It does so by emphasising the importance of identifying the nature of the energy levels which are involved in swapping quanta of energy with electromagnetic radiation. Transitions take place by the atom or molecule interacting with the electric or magnetic field of the radiation. For this to happen successfully the fields must encounter an electric or magnetic dipole moment. In some cases we saw how the selection rules that govern the occurrence of spectral lines could be related to the demand that the transition dipole moment be non-zero.

For e.s.r. and n.m.r. we did not examine much of the theory of how the transitions take place. The reason is that to understand the mathematics we must develop some new skills and insights into quantum theory. This is the task of the next chapter.

M.5.1

We can illustrate the validity of the selection rule $\Delta l = \pm 1$ in the following way. If we assume that the electric field is confined to the z-axis, then

$$d_z = -e \int \psi_2 z \psi_1 \mathrm{d}v.$$

But $z = r \cos \theta$ so

$$d_z = -e \int \psi_2 r \cos \theta \psi_1 \mathrm{d}v$$

$$= -e \int_0^{2\pi} \int_0^\pi \int_0^\infty \psi_2 r \cos \theta \psi_1 r^2 \mathrm{d}r \mathrm{d}\theta \mathrm{d}\phi.$$

If ψ_1 is a 1s orbital and ψ_2 a 2s orbital of hydrogen,

$$d_z = -e \int_0^{2\pi} \int_0^\pi \int_0^\infty \frac{1}{4\sqrt{(2\pi)}} \left(\frac{1}{a_0}\right)^{3/2}$$

$$\times (2 - r/a_0) e^{-r/2a_0} r \cos \theta$$

$$\times \frac{1}{\sqrt{\pi}} \left(\frac{1}{a_0}\right)^{3/2} e^{-r/a_0} r^2 \mathrm{d}r \mathrm{d}\theta \mathrm{d}\phi$$

$$= -\frac{e}{4\pi\sqrt{2}} \left(\frac{1}{a_0}\right)^3 \int_0^\infty (2 - r/a_0) e^{-r/2a_0}$$

$$\times r e^{-r/2a_0} r \mathrm{d}r \int_0^\pi \cos \theta \mathrm{d}\theta \int_0^{2\pi} \mathrm{d}\phi.$$

But, $\int_0^\pi \cos \theta \mathrm{d}\theta = 0$ so $d_z = 0$ and we say that the transition is forbidden. This agrees with our selection rule because a transition be- tween s orbitals has $\Delta l = 0$.

Now look at a transition between a 1s and 2p$_z$ orbital. Simplifying the terms,

$$d_z = -\frac{e}{4\pi\sqrt{2}} \left(\frac{1}{a_0}\right)^4 \int_0^\infty r^4 e^{-3r/2a_0} \mathrm{d}r$$

$$\times \int_0^\pi \cos^2 \theta \mathrm{d}\theta \int_0^{2\pi} \mathrm{d}\phi.$$

Using the standard integrals in appendix A, we find

$$d_z = -\frac{e}{4\pi\sqrt{2}} \left(\frac{1}{a_0}\right)^4 \left[\frac{4!}{(3/2a_0)^5}\right]$$

$$\times \left[\frac{1}{2}(\theta + \frac{1}{2}\sin 2\theta)\right]_0^\pi \times 2\pi$$

$$= -\frac{64\pi}{81\sqrt{2}} e a_0$$

which is clearly non-zero. We have therefore determined the value of the transition dipole moment for a transition for which $\Delta l = +1$.

M.5.2

A particle displaying simple harmonic motion suffers a restoring force, F, proportional to its displacement from its equilibrium position, i.e.

$$F = -kx$$

where x is the displacement and k, the constant of proportionality, is the force constant. Because potential energy is related to the force by $F = -dV/dx$ it follows that

$$dV = -Fdx$$

or

$$V = -\int Fdx.$$

Here

$$V = k \int_0^x xdx$$

or

$$V = \tfrac{1}{2}kx^2.$$

This relation explains the shape of the curve of fig. 5.20. Thus, the total energy, E, is

$$E = \frac{p^2}{2m} + \tfrac{1}{2}kx^2.$$

This can easily be converted into the form required for the Schrödinger equation as

$$-\frac{\hbar^2}{2m}\frac{\partial^2 \psi}{\partial x^2} + \tfrac{1}{2}kx^2\psi = E\psi.$$

We shall not go through the mathematics needed to solve the equation, but the solutions for the wavefunctions can be written in terms of a series of polynomials called Hermite polynomials. (see table 5.3). The wavefunctions drawn in fig. 5.20 show the form of the first three Hermite polynomials.

If we write the force, F, above as $m(d^2x/dt^2)$ then

$$m\frac{d^2x}{dt^2} = -kx.$$

This has the solution $x = A\sin(2\pi f_0 t)$, where, by substituting into the equation we can derive the relation $f_0 = (1/2\pi)\sqrt{(k/m)}$.

If a molecule vibrates with simple harmonic motion then we can decrease the difficulty of analysing the motion by dealing with the reduced mass instead of the individual masses. For two atoms of masses m_1 and m_2, the reduced mass, μ, is

$$\mu = \frac{m_1 m_2}{m_1 + m_2}.$$

The corresponding formula for the frequency is

$$f_0 = \frac{1}{2\pi}\sqrt{\left(\frac{k}{\mu}\right)}.$$

M.5.3

The kinetic energy of a single particle rotating in a plane about a central axis, z, is given by

$$E = \frac{L_z^2}{2I}$$

where L_z is the angular moment about the z-axis and $I = mr^2$ is the moment of inertia. Just as in the case of vibrations of a diatomic molecule, it is possible to write down an analogous equation for the rotations of such a molecule where this time $I = \mu r^2$ with μ the reduced mass. This is the classical expression for the energy. As we should expect now, quantum theory says that we must obtain the value of the energy from the corresponding wavefunction of the system. For linear kinetic energy we saw that terms like $p_x^2/2m$ in classical theory re-appeared in the Schrödinger equation as $-(\hbar^2/2m)(\partial^2/\partial x^2)$. That is, as a differential operator. The differential operator corresponding to L_z^2 is $-\hbar^2(\partial^2/\partial \phi^2)$ where ϕ represents the azimuthal angle (see fig. 5.29). Thus, in quantum theory, the differential operator which will eventually give us the energy of a plane rigid rotator is

$-(\hbar^2/2I)(\partial^2/\partial\phi^2)$. (There is an obvious similarity between this and the corresponding expression for the linear kinetic energy.) If we call the wavefunction for the rotator $\psi(\phi)$ we have

$$-\frac{\hbar^2}{2I}\frac{\partial^2[\psi(\phi)]}{\partial\phi^2} = E\psi(\phi).$$

To solve this equation we should first note that $\psi(\phi)$ must be periodic for ϕ repeats every 2π. Therefore it is natural to seek solutions involving trigonometric functions. This guarantees that we are going to obtain quantised solutions.

We shall choose the solution

$$\psi(\phi) = A\cos(m\phi)$$

where m is an integer and A is the amplitude. Then

$$-\frac{\hbar^2}{2I}\frac{\partial^2[\psi(\phi)]}{\partial\phi^2} = \frac{\hbar^2}{2I}\frac{\partial[-mA\sin(m\phi)]}{\partial\phi}$$

$$= \frac{\hbar^2}{2I}m^2 A\cos(m\phi)$$

$$= \frac{m^2\hbar^2}{2I}\psi(\phi).$$

Thus we find that

$$E_m = \frac{m^2\hbar^2}{2I}.$$

Notice that the appearance of the square of the quantum number, m, means that the energy levels are doubly degenerate, e.g. $m = +1$ and $m = -1$ give identical energies. The formula for E_m also shows why the spacing of rotational levels increases with increasing energy.

M.5.4

Suppose a collection of molecules can exist in two different energy states, E_0, E_1 with E_1 higher than E_0. The populations of each state n_0, n_1 are related by the Boltzmann equation

$$n_1 = n_0 e^{-\Delta E/kT}$$

where ΔE is the difference in energy between E_1 and E_0; k is the Boltzmann constant, $1.38 \times 10^{-23}\,\mathrm{J\,K^{-1}}$, and T the absolute temperature. This relation only holds if the energy levels are not degenerate. If, for example, E_1 has a degeneracy g we find that

$$n_1 = gn_0 e^{-\Delta E/kT}.$$

We have already seen that the solutions of the Schrödinger equation for the hydrogen atom are $2l + 1$ degenerate. Similarly, the rotational energy levels are $2J + 1$ degenerate. Thus, for rotational energy levels,

$$n_J = (2J + 1)n_0 e^{-BJ(J+1)/kT}.$$

It so happens that the values of B and J conspire to make n_J a maximum when J is about two or three for carbon monoxide. Because there are more molecules in these rotational states than any other, it follows that there will be more transitions involving these states than any other.

The Boltzmann distribution has far reaching implications in spectroscopy and thermodynamics. In general, the separation between electronic energy levels and between vibrational levels is very large compared to kT at room temperature. This means that only the ground state levels are occupied. However, if the ground state electrons are excited to an upper level faster than they can return downwards there is an inversion of the Boltzmann population. This is the basis of the laser. The trick of finding a good laser is to use a system that traps the electrons in an upper state for a sufficiently long time that eventually a very large number of transitions can occur from an upper level to the ground state. The power developed when a large number of such transitions occur together can be huge, of the order of megawatts.

Part 2
Taking it further

6

Formal quantum theory

6.1 Introduction

We have seen that quantum theory can account for a wide range of properties of atoms and molecules. It is possible to become so involved with applying the theory that the meaning and importance of the ideas contained in quantum theory become obscured or, at least, taken for granted. In order to avoid these pitfalls we shall go below the surface structure and successes of quantum theory and delve into its deep structure.

There is no escape from the fact that quantum theory is fundamentally mathematical in its content. Until now we have put much of the mathematics to one side, allowing it to take second place to the ideas. In this chapter we shall find that the mathematics and the ideas become increasingly intertwined. There are three themes which we will find particularly useful to develop. They are complex numbers, vector algebra, and the theory of operators. They form the content of the next three sections.

6.2 Complex numbers

Much of a chemist's time is taken up with reading values given by a variety of measuring instruments such as pH meters, thermometers, or spectrometers. These values are always real numbers, or scalars. Any observable quantity – observable for short – has to be a real number. However, quantities which are not observable need not be real numbers. They can be complex numbers. Although it is not mathematically rigorous, by complex number we shall mean a number which contains the square root of minus one, $\sqrt{(-1)}$. This we will give the symbol i. Although i has no direct physical significance, it is by no means useless.

For example, the complex exponentials $e^{i\theta}$ and $e^{-i\theta}$ serve to define the real trigonometric function $\cos\theta$ and $\sin\theta$:

$$\cos\theta = \tfrac{1}{2}(e^{i\theta} + e^{-i\theta})$$

$$\sin\theta - \frac{1}{2i}(e^{i\theta} - e^{-i\theta}).$$

Complex numbers usually consist of a real part and an imaginary part. If we write the complex number c as

$$c = a + ib$$

then a is the real part. In spite of the fact that a and b are both real numbers, b is called the imaginary part. The complex conjugate of c is written as c^* and is obtained by replacing i by $-$ i whenever it appears. Thus

$$c^* = (a + ib)^* = a - ib.$$

By combining our definitions of c and c^* we find that the real part of c is

$$a = \tfrac{1}{2}(c + c^*).$$

Writing this real part as Re c we have

$$\text{Re } c = \tfrac{1}{2}(c + c^*).$$

Similarly, the imaginary part, b is Im c where

$$\text{Im } c = \frac{1}{2i}(c - c^*).$$

Interestingly, we see that

$$\cos\theta = \text{Re } e^{i\theta}$$

and

$$\sin\theta = \text{Im } e^{i\theta}.$$

Because of this fairly simple relation between $\cos\theta$, $\sin\theta$ and $e^{i\theta}$ it is often found just as useful to work with their complex exponential form as with $\cos\theta$ and $\sin\theta$ themselves. Indeed, the angular wavefunctions of hydrogen in table 3.4 do not actually occur directly as solutions of Schrödinger's equation. The initial solutions for the angular wavefunctions involve a part which depends on the aximuthal angle,

ϕ, and the quantum number m_l in the form

$$\Phi(\phi) = \frac{1}{\sqrt{(2\pi)}} e^{im_l\phi}.$$

For an s orbital, $l = 0$ so $m_l = 0$ and the factor $1/\sqrt{(2\pi)}$ only remains. For p orbitals $l = 1$ so m_l can be $+1$, 0, -1. We obtain the corresponding wavefunction

$$\Phi_1(\phi) = \frac{1}{\sqrt{(2\pi)}} e^{i\phi}$$

$$\Phi_0(\phi) = \frac{1}{\sqrt{(2\pi)}}$$

$$\Phi_{-1}(\phi) = \frac{1}{\sqrt{(2\pi)}} e^{-i\phi}.$$

If you look at table 3.4 you will see that when $m_l = \pm 1$ we find a factor $\cos\phi$ or $\sin\phi$. Many people prefer working with these real functions, partly because they can be more easily used to plot graphs. The principle of superposition was applied to form the combinations $(1/\sqrt{2})[\Phi_1(\phi) + \Phi_{-1}(\phi)]$ and $(1/i\sqrt{2})[\Phi_1(\phi) - \Phi_{-1}(\phi)]$. (Check back to chapter 2 if you have forgotten about this principle.) The former combination gives the $\cos\phi$ term, the latter the $\sin\phi$ term.

In practice it is not at all uncommon for wavefunctions to be given as complex numbers. At first sight there seems to be a problem here because we have already said that observables must be real numbers. But this is precisely the point! We found in chapter 2 that according to Born the wavefunction itself tells us nothing of physical significance. The wavefunction is quite entitled to be complex, but its square must not be complex because it corresponds to a probability density function built up from real numbers.

Suppose we have a complex number $c = a + ib$. If we form the product $c \times c$ we find

$$c \times c = (a + ib)(a + ib) = a^2 - b^2 + 2iab$$

which is still a complex number. However, the product c^*c gives

$$c^*c = (a - ib)(a + ib) = a^2 + b^2$$

which is purely real.

For this reason we define the square of the magnitude of a complex wavefunction, ψ, as $\psi^*\psi$.

This product we wrote as $|\psi|^2$ in previous chapters. In future we must be careful to note that $|\psi|^2 \equiv \psi^*\psi$.

One result of this is that our condition for normalisation is now

$$\int \psi^*\psi \, dv = 1.$$

We have seen that a wavefunction, e.g. a molecular orbital, may be built up by combining other wavefunctions, e.g. atomic orbitals. A combination $\psi = c_1\phi_1 + c_2\phi_2 + c_3\phi_3 + \cdots + c_n\phi_n$ means that

$$\int (c_1\phi_1 + c_2\phi_2 + c_3\phi_3 + \cdots + c_n\phi_n)^*$$

$$\times (c_1\phi_1 + c_2\phi_2 + c_3\phi_3 + \cdots + c_n\phi_n) dv = 1$$

i.e.

$$c_1^*c_1 \int \phi_1^*\phi_1 \, dv + c_2^*c_2 \int \phi_2^*\phi_2 \, dv$$

$$+ c_3^*c_3 \int \phi_3^*\phi_3 \, dv + \cdots + c_n^*c_n \int \phi_n^*\phi_n \, dv$$

$$+ c_1^*c_2 \int \phi_1^*\phi_2 \, dv + c_1^*c_3 \int \phi_1^*\phi_3 \, dv$$

$$+ \cdots = 1. \tag{A}$$

There are two types of integral here, which we can write as $\int \phi_i^*\phi_i \, dv$ and $\int \phi_i^*\phi_j \, dv$. If these wavefunctions form a normalised and orthogonal set,

$$\int \phi_i^*\phi_i \, dv = 1$$

while

$$\int \phi_i^*\phi_j \, dv = 0. \qquad (i \neq j).$$

We can summarise these conditions in one equation:

$$\int \phi_i^*\phi_j \, dv = \delta_{i,j}.$$

$\delta_{i,j}$ is the Kronecker delta. It is defined by the relations

$$\delta_{i,j} = 1 \qquad \text{if} \qquad i = j,$$

$$\delta_{i,j} = 0 \qquad \text{if} \qquad i \neq j.$$

For example, all the terms in equation (A) are covered by the terms in the expansion of

$$\sum \sum c_i^*c_j \int \phi_i^*\phi_j \, dv.$$

But this is just $\sum \sum c_i^*c_j\delta_{i,j}$ and

$$\sum \sum c_i^*c_j\delta_{i,j} = \sum c_i^*c_i$$

$$= c_1^*c_1 + c_2^*c_2$$

$$+ c_3^*c_3 + \cdots + c_n^*c_n.$$

Shortly we shall see the significance of this.

Questions

6.1 In this question, use $c = a + ib$ and $d = e + if$. Simplify (i) d^*d^*; (ii) Im (ic); (iii) $(\cos\theta + i\sin\theta)^n$; (iv) Re $(cd^* + c^*d)$; (v) Re $(cd^* - c^*d)$.

6.2 You should know that $\int_a^b e^{kx} dx = [(1/k)e^x]_a^b$. If n is an integer,
   (i) what is $\int_0^{2\pi} e^{in\phi} d\phi$ when $n = 0$?
   (ii) what is $\int_0^{2\pi} e^{in\phi} d\phi$ when $n \neq 0$?

6.3 If we return to section 5.5, we find that the energy of a plane rigid rotator is found by solving the equation

$$-\frac{\hbar^2}{2I} \frac{\partial^2 \psi_m(\phi)}{\partial\phi^2} = E\psi_m(\phi).$$

   (i) Show by differentiating that $\psi_m(\phi) = Ae^{im\phi}$ is a solution of the equation with $E = m^2\hbar^2/2I$.
   (ii) For a wavefunction to be normalised we have required

$$\int |\psi|^2 dv = 1.$$

   For complex wavefunctions, this means

$$\int \psi^*\psi \, dv = 1.$$

   In the case of the plane rigid rotator, we do not have to integrate over all space as there is only one variable, the angle ϕ. So,

$$\int_0^{2\pi} \psi_m^*(\phi)\psi_m(\phi) d\phi = 1.$$

   With $\psi_m(\phi) = Ae^{im\phi}$, evaluate this integral and calculate the value of the normalising constant, A. Hint: see your answer to question 6.2 (i).

6.3 Vectors

If we wish to describe the position of a point in three dimensional space by a vector, **r**, we choose to write **r** as a combination of three unit vectors **i**, **j**, **k** oriented along the x-, y-, and z-axes. The scalar products between the unit vectors have the property that

$$\mathbf{i}\cdot\mathbf{i}=\mathbf{j}\cdot\mathbf{j}=\mathbf{k}\cdot\mathbf{k}=1,$$
$$\mathbf{i}\cdot\mathbf{j}=\mathbf{j}\cdot\mathbf{i}=\mathbf{i}\cdot\mathbf{k}=\mathbf{k}\cdot\mathbf{i}=\mathbf{j}\cdot\mathbf{k}=\mathbf{k}\cdot\mathbf{j}=0.$$

The general vector equation for any point in three-dimensional space is

$$\mathbf{r}=c_1\mathbf{i}+c_2\mathbf{j}+c_3\mathbf{k}.$$

The coefficients c_1, c_2, and c_3 give the projections of **r** on the x-, y-, and z-axes. If we wish to isolate the coefficient c_1 we form the scalar product of **r** with **i**:

$$\mathbf{i}\cdot\mathbf{r}=c_1\mathbf{i}\cdot\mathbf{i}+c_2\mathbf{i}\cdot\mathbf{j}+c_3\mathbf{i}\cdot\mathbf{k}$$

or,

$$\mathbf{i}\cdot\mathbf{r}=c_1.$$

The unit vector **i** is said to project **r** onto the x-axis. Similarly **j** and **k** will project **r** onto the y- and z-axes (fig. 6.1).

Two vectors, $\mathbf{r}_1$ and $\mathbf{r}_2$, have their scalar product defined by

$$\mathbf{r}_1\cdot\mathbf{r}_2=|\mathbf{r}_1||\mathbf{r}_2|\cos\theta$$

where $|\mathbf{r}_1|$ and $|\mathbf{r}_2|$ are the magnitudes, or lengths of the two vectors and θ is the angle between them. When $\theta=0$, $\mathbf{r}_1$ and $\mathbf{r}_2$ are parallel, but if $\theta=\pi$ they are antiparallel. When $\theta=\pi/2$ they are perpendicular to one another. Alternatively, we say that they are orthogonal.

The length of the vector **r** in fig. 6.1 is $\sqrt{(c_1{}^2+c_2{}^2+c_3{}^2)}$. If a vector has unit length it is normalised. It is easy to normalise any vector such as **r** by dividing by its length. Thus

$$\mathbf{r}'=\frac{1}{\sqrt{(c_1{}^2+c_2{}^2+c_3{}^2)}}(c_1\mathbf{i}+c_2\mathbf{j}+c_3\mathbf{k})$$

is normalised.

We know that any vector in three-dimensional space can be built up from a combination of the basic vectors **i**, **j**, and **k**. This set of three unit vectors is known as a complete set of vectors. One less would restrict us to two dimensions, one more would take us into four dimensions. Initially, the notion of vectors in more than three dimensions may seem odd, but this is only because we cannot visualise them all, or draw pictures of them. It is quite simple to go into four dimensions, all we do is invent a fourth unit vector, **l**, say. Then we can write vectors like

$$\mathbf{r}_1=\frac{1}{\sqrt{54}}(2\mathbf{i}+3\mathbf{j}+4\mathbf{k}+5\mathbf{l})$$

and

$$\mathbf{r}_2=\frac{1}{\sqrt{54}}(5\mathbf{i}+4\mathbf{j}-3\mathbf{k}-2\mathbf{l})$$

and we find that

$$\begin{aligned}\mathbf{r}_1\cdot\mathbf{r}_2&=\tfrac{1}{54}(2\mathbf{i}+3\mathbf{j}+4\mathbf{k}+5\mathbf{l})\\&\quad\cdot(5\mathbf{i}+4\mathbf{j}-3\mathbf{k}-2\mathbf{l})\\&=\tfrac{1}{54}(2\times5+3\times4-4\times3-5\times2)\\&=0.\end{aligned}$$

Therefore $\mathbf{r}_1$ and $\mathbf{r}_2$ are orthogonal. They also happen to be normalised.

We can extend the argument to five, six, seven,... even to n dimensions. To do so, it is more convenient to use a different notation for our base vectors. Just as **i**, **j**, **k** and **i**, **j**, **k**, **l** form complete sets for three- and four-dimensional space, we can call the n basic vectors for n-dimensional space $\mathbf{v}_1$, $\mathbf{v}_2$, $\mathbf{v}_3$, $\mathbf{v}_4$,..., $\mathbf{v}_n$. This set, $\{\mathbf{v}_i\}$ for short, is said to span n-dimensional space. A general way of writing a vector **V** in this space is to put

$$\mathbf{V}=c_1\mathbf{v}_1+c_2\mathbf{v}_2+c_3\mathbf{v}_3+\cdots+c_n\mathbf{v}_n$$

Fig. 6.1. The vector $\mathbf{r}=c_1\mathbf{i}+c_2\mathbf{j}+c_3\mathbf{k}$ has projections c_1, c_2, c_3 on the x-, y- and z-axes respectively.

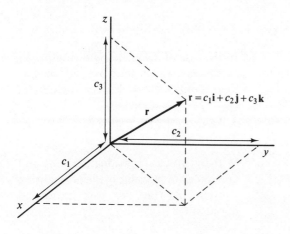

or

$$V = \sum_i c_i v_i.$$

The requirement that

$$v_1 \cdot v_1 = v_2 \cdot v_2 = \cdots = v_i \cdot v_i = 1$$

while

$$v_1 \cdot v_2 = v_1 \cdot v_3 = \cdots = v_2 \cdot v_1$$
$$= v_2 \cdot v_3 = \cdots = v_i \cdot v_j = 0$$

is summarised by writing

$$v_i \cdot v_j = \delta_{i,j}$$

where $\delta_{i,j}$ is the Kronecker delta.

For example,

$$|V|^2 = \left(\sum_i c_i v_i\right) \cdot \left(\sum_j c_j v_j\right)$$
$$= \sum_i \sum_j c_i c_j v_i \cdot v_j$$
$$= \sum_i \sum_j c_i c_j \delta_{i,j}$$
$$= \sum_i c_i c_i \quad \text{(or } \sum_j c_j c_j \text{ – the choice of}$$
$$\text{subscript is irrelevant)}$$
$$= c_1{}^2 + c_2{}^2 + c_3{}^2 + \cdots + c_n{}^2.$$

We have assumed that all these coefficients and vectors are real, but this is not essential. Let us define two new unit vectors which we call v_1 and v_2. Suppose they have the properties that $v_1 \cdot v_2 = 0$, $v_1 \cdot v_1 = 1$, but $v_2 \cdot v_2 = -1$. The first two relations are 'normal' but if v_2 is purely real the third relation is impossible. If v_2 is complex it is quite possible. If we assume that v_2 is complex we should put $v_2{}^* \cdot v_2 = 1$. The vectors v_1 and v_2 will span a two-dimensional, complex, vector space. We can visualise this behaviour if by way of example we put $v_1 \equiv i$ and $v_2 \equiv ij$. The vector $V_1 = v_1 + v_2$ takes us to the point P_1 on the Argand diagram of fig. 6.2. Similarly $V_2 = v_1 - v_2$ takes us to the point P_2. From fig. 6.2 V_1 and V_2 have the same length, $\sqrt{2}$. But if

$$|V_1|^2 = V_1 \cdot V_1$$
$$= (v_1 + v_2) \cdot (v_1 + v_2)$$
$$= v_1 \cdot v_1 + v_1 \cdot v_2 + v_2 \cdot v_1 + v_2 \cdot v_2$$
$$= 1 + 0 + 0 - 1,$$

so $|V_1|^2 = 0.$

The result is obviously wrong. However we met a

similar state of affairs in the previous section. The remedy is to define

$$|V_1|^2 = V_1{}^* \cdot V_1.$$

Then

$$|V_1|^2 = (v_1 + v_2)^* \cdot (v_1 + v_2)$$
$$= v_1{}^* \cdot v_1 + v_1{}^* \cdot v_2 + v_2{}^* \cdot v_1 + v_2{}^* \cdot v_2$$
$$= 1 + 0 + 0 + 1.$$

(Check these for yourself by writing i for v_1 and ij for v_2.) Then,

$$|V_1| = \sqrt{2}.$$

Likewise,

$$|V_2| = \sqrt{2}.$$

We could write the general vector in the Argand diagram as

$$V = c_1 v_1 + c_2 v_2$$

where the coefficients c_1 and c_2 can also be complex. If we go to a three-dimensional complex space,

$$V = c_1 v_1 + c_2 v_2 + c_3 v_3.$$

Fig. 6.2. The two points P_1, P_2 on an Argand diagram have position vectors $v_1 + v_2$, $v_1 - v_2$. The lengths of OP_1 and OP_2 are both $\sqrt{2}$.

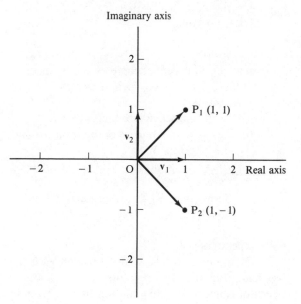

Generalising to n-dimensions,

$$\mathbf{V} = \sum c_i \mathbf{v}_i$$

but now,

$$|\mathbf{V}|^2 = \left(\sum_i c_i \mathbf{v}_i\right)^* \cdot \left(\sum_j c_j \mathbf{v}_j\right)$$

$$= \sum_i \sum_j c_i^* c_j \mathbf{v}_i^* \cdot \mathbf{v}_j$$

$$= \sum_i \sum_j c_i^* c_j \delta_{i,j}$$

$$= \sum_i c_i^* c_i$$

(or $\sum c_j^* c_j$, the choice of subscript does not matter because in either case, $|\mathbf{V}|^2 = c_1^* c_1 + c_2^* c_2 + \cdots + c_n^* c_n$). We used the relation

$$\mathbf{v}_i^* \cdot \mathbf{v}_j = \delta_{i,j}$$

which, for complex vectors, is the equivalent of $\mathbf{v}_i \cdot \mathbf{v}_j = \delta_{i,j}$ for real vectors.

Starting with the common unit vectors spanning a three-dimensional, real, vector space, in a few pages we have developed sufficient knowledge to cope with forming vectors from unit vectors spanning n-dimensional, complex, vector space.

Questions

6.4 (i) Normalise the vector $\mathbf{r} = 2\mathbf{i} + 4\mathbf{j} + 6\mathbf{k}$.
(ii) Find a vector which is orthogonal to it.

6.5 Are the vectors $\mathbf{r}_1 = 3\mathbf{i} + 7\mathbf{j} - 5\mathbf{k}$, $\mathbf{r}_2 = 45\mathbf{i} + 27\mathbf{k}$ orthogonal?

6.6 The vector $\mathbf{V}_1 = 2i\mathbf{v}_1 + 3\mathbf{v}_2 + 4\mathbf{v}_3 + 5i\mathbf{v}_4$ exists in a four dimensional complex vector space in which $\{\mathbf{v}_1, \mathbf{v}_2, \mathbf{v}_3, \mathbf{v}_4\}$ form a complete set of unit vectors.
(i) Form the scalar product $\mathbf{V}_1^* \cdot \mathbf{V}_1$.
(ii) Take the square root of the result to find the length of $\mathbf{V}_1$.
(iii) Hence write down $\mathbf{V}_1$ in its normalised form.
(iv) Show that $\mathbf{V}_1$ is orthogonal to $\mathbf{V}_2 = 6i\mathbf{v}_1 + 2\mathbf{v}_2 + 3\mathbf{v}_3 - 6i\mathbf{v}_4$.

6.4 Operators

Quantum theory is riddled through with the use and properties of operators. The idea of an operator is not unique to quantum theory. Even in elementary mathematics operators are used, but they are not always called by this name. For example, the expression $\sqrt{25}$ means 'take the square root of 25'. The sign $\sqrt{}$ is an instruction to carry out a particular mathematical operation, so it acts as a simple operator. We can create a symbol for other operators. For example, if

$$S(5) = 25$$

and

$$S(-7) = 49$$

we recognise S as the operator for squaring a number.

The calculus is based on the use of differential operators such as d/dx. The rule is that when d/dx acts as x^n we find

$$\frac{d(x^n)}{dx} = nx^{n-1}.$$

The operation of differentiation can be repeated. Thus

$$\frac{d^2(x^n)}{dx^2} = \frac{d}{dx}\frac{d(x^n)}{dx}$$

$$= \frac{d(nx^{n-1})}{dx}$$

$$= n(n-1)x^{n-2}.$$

For functions of more than one variable, we use partial differential operators like $\partial/\partial x$. For example,

$$\frac{\partial^2(x^2 y^3 z^4)}{\partial y^2} = \frac{\partial}{\partial y}\frac{\partial(x^2 y^3 z^4)}{\partial y}$$

$$= \frac{\partial(3x^2 y^2 z^4)}{\partial y}$$

$$= 6x^2 y z^4$$

because $\partial/\partial y$ only acts on y^n.

An important partial differential operator which we have met before, section 2.2, is the Laplacian, ∇^2:

$$\nabla^2 \equiv \frac{\partial^2}{\partial x^2} + \frac{\partial^2}{\partial y^2} + \frac{\partial^2}{\partial z^2}.$$

Some operators and the functions they operate on behave in a particularly useful way. Sometimes we find that

operator × function = a number × the same

function.

When this happens, the function is said to be an eigenfunction of the operator. The number it yields is its eigenvalue. Then,

an operator × eigenfunction

= eigenvalue × eigenfunction.

Very often eigenfunctions of differential operators appear in quantum theory. For example,

$$\frac{d(e^{kx})}{dx} = ke^{kx}.$$

Thus e^{kx} is an eigenfunction of d/dx with eigenvalue k.

It is often useful to write equations such as this in a symbolic form. For example, with $\hat{D} \equiv d/dx$ and $f(x) \equiv e^{kx}$ we have

$$\hat{D}f(x) = kf(x).$$

We will see shortly that Schrodinger's equation is an eigenvalue equation which can be written in symbolic form. The 'hat', ^, on a symbol, like that on the D, will be used to indicate an operator.

An operator, let us call it $\hat{A}$, might act on a combination of functions, f_1 and f_2. If

$$\hat{A}(f_1 + f_2) = \hat{A}f_1 + \hat{A}f_2$$

we say that $\hat{A}$ is a linear operator. This might seem trivial unless we think of some operators, like $\sqrt{\ }$, or taking the logarithm, which certainly are not linear. For instance,

$$\sqrt{(9 + 16)} \neq \sqrt{9} + \sqrt{16}.$$

A feature of the operators in quantum theory is that they are all linear.

We shall investigate another property of operators in the following way. A very simple operation on a function is to multiply it by a number. Rather more involved is the operation of differentiation. Suppose we operate on x^2 with x and with d/dx.

$$x\frac{d(x^2)}{dx} = x \cdot 2x$$

$$= 2x^2.$$

But, if we do this in the reverse order,

$$\frac{d(xx^2)}{dx} = \frac{d(x^3)}{dx}$$

$$= 3x^2.$$

Thus,

$$\left(x\frac{d}{dx} - \frac{d}{dx}x \right)x^2 \neq 0,$$

and, in general

$$x\frac{d}{dx}f(x) - \frac{d}{dx}xf(x) \neq 0,$$

or

$$\left(x\frac{d}{dx} - \frac{d}{dx}x \right)f(x) \neq 0.$$

However, taking d/dx and d^2/dx^2 as our pair of operators,

$$\frac{d}{dx}\frac{d^2(x^2)}{dx^2} = \frac{d}{dx}\frac{d}{dx}\frac{d(x^2)}{dx}$$

$$= \frac{d}{dx}\frac{d(2x)}{dx}$$

$$= \frac{d(2)}{dx}$$

$$= 0.$$

And

$$\frac{d^2}{dx^2}\frac{d(x^2)}{dx} = \frac{d}{dx}\frac{d(2x)}{dx}$$

$$= 0.$$

as before. Thus

$$\left(\frac{d}{dx}\frac{d^2}{dx^2} - \frac{d^2}{dx^2}\frac{d}{dx} \right)x^2 = 0.$$

You might like to prove that these relations between the operators we have used are quite general. That is, for any function $f(x)$,

$$\left(x\frac{d}{dx} - \frac{d}{dx}x \right)f(x) \neq 0$$

and

$$\left(\frac{d}{dx}\frac{d^2}{dx^2} - \frac{d^2}{dx^2}\frac{d}{dx} \right)f(x) = 0$$

Generalising these results, for any two operators $\hat{A}$ and $\hat{B}$, either

$$(\hat{A}\hat{B} - \hat{B}\hat{A})f(x) = 0$$

or

$$(\hat{A}\hat{B} - \hat{B}\hat{A})f(x) \neq 0.$$

In the first instance, $\hat{A}$ and $\hat{B}$ are said to commute. In the second instance, they do not commute. Because of the importance of combinations of operators we shall use a special notation. We shall write the combination $(\hat{A}\hat{B} - \hat{B}\hat{A})$ as $[\hat{A}, \hat{B}]$. The bracket $[\hat{A}, \hat{B}]$ is a commutator bracket. If $\hat{A}$ and $\hat{B}$ commute we write this as

$$[\hat{A}, \hat{B}] = 0.$$

This is a symbolic way of writing the relationship between $\hat{A}$ and $\hat{B}$. To prove whether two operators commute, they have to operate on something. This point is illustrated in the questions that follow. Commuting operators take pride of place in quantum theory because they are the operators which give us the most information about a system. When operators do not commute we shall find that there is a restriction on the information they can give us. This restriction is enshrined in Heisenberg's uncertainty principle.

Questions

6.7 Which of the functions, x^2, e^{kx}, e^{ikx} are eigenfunctions of (i) d/dx, (ii) d^2/dx^2, and what are their eigenvalues?

6.8 Prove that $[x(d/dx) - (d/dx)x]f(x) = -f(x)$ for any $f(x)$.

6.9 Do x and d^2/dx^2 commute? To answer this, form the commutator bracket $[x, d^2/dx^2]$ and operate with it on a function $f(x)$. You will find it helpful to write $(d/dx)f(x) = f'(x)$ and $(d^2/dx^2)f(x) = f''(x)$. Note: don't forget to differentiate a product correctly.

6.10 A particularly important set of operators in quantum theory are the angular momentum operators. They are defined by

$$\hat{L}_x = \frac{\hbar}{i}\left(y\frac{\partial}{\partial z} - z\frac{\partial}{\partial y}\right)$$

$$\hat{L}_y = \frac{\hbar}{i}\left(z\frac{\partial}{\partial x} - x\frac{\partial}{\partial z}\right)$$

$$\hat{L}_z = \frac{\hbar}{i}\left(x\frac{\partial}{\partial y} - y\frac{\partial}{\partial x}\right)$$

$$\hat{\mathbf{L}}^2 = \hat{L}_x^{\,2} + \hat{L}_y^{\,2} + \hat{L}_z^{\,2}.$$

By operating on a function, which you can write as F for short, prove that

(i) $[\hat{L}_x, y] = i\hbar z$

(ii) $[\hat{L}_x, \hat{L}_y] = i\hbar \hat{L}_z$

(iii) $[\hat{L}_z^{\,2}, \hat{L}_z] = 0$

(iv) $[\hat{L}_z, r^2] = 0$ where $r^2 = x^2 + y^2 + z^2$

(v) $[\hat{\mathbf{L}}^2, \hat{L}_z] = 0.$

Remember, again, to differentiate a product correctly and note that terms like $\partial y/\partial x$ or $\partial z/\partial y$ are zero. Why?

Can you generalise the result in (ii) to predict the outcome of $[\hat{L}_y, \hat{L}_z]$ and $[\hat{L}_z, \hat{L}_x]$?

6.11 If $\hat{p}_x = (\hbar/i)(\partial/\partial x)$, prove that $[\hat{p}, x] = \hbar/i$.

6.5 States

A primary aim of classical and quantum theory is to predict successfully the results of experiments. In order to make such predictions both theories employ the idea of the state of a system. Ideally, contained within this concept is all the information about the system which is to provide a basis upon which to work out the predictions. This notion probably appears rather vague but an example may clarify the issue, at least as far as classical theory is concerned – it is not so easy to achieve a similar aim for quantum physics and we shall leave this for a moment. The example we shall use is the freely moving particle.

Suppose a particle of mass m is moving freely in space, which means that it is moving in the absence of any forces. A typical classical problem would be to predict the position, momentum, and energy of the particle at any time. In order to make predictions it will be necessary to make some initial observations in order to fix some reference points from which to measure the fundamental quantities of length and time. That is, an observation of the particle will have to be made at a measured time, t_0, and at that time its position measured as x_0, say. There is one further value which is required: the velocity of the particle at t_0, which will be called u. Armed with these three pieces of information together with Newton's laws, the required predictions can be made. Thus, because there are no forces acting on the particle it will continue to travel in a straight line with no change in its velocity. Therefore,

- (i) its velocity is u at all later times.
- (ii) its momentum, p, will always be $p = mu$,
- (iii) its energy, E, will always be $E = \frac{1}{2}mu^2$, or $E = p^2/2m$,
- (iv) its position, x_1, at a later time, t_1, will be given by the equation $x_1 - x_0 = u(t_1 - t_0)$.

All the required predictions about the state of the particle at any time can therefore be made following initial measurements of time, position, and velocity. These three quantities define the initial state of the system. Notice that the 'things' (t_0, x_0, and u) which specify the state are values of observables.

In quantum theory the classical ideas of the state and of our ability to measure the observables is changed radically in many respects. To come to an understanding of the differences we shall make use of a method of dealing with quantum states that was invented by P.A.M. Dirac. His method relies on using the mathematics of vector algebra and operators. The vectors he used span a particular species of complex vector space called Hilbert space. (Hilbert was a German pure mathematician who became heavily involved in the ramifications of Schrödinger's work.)

Dirac proposed a novel notation for his vectors. He suggested that, whatever the detailed nature of a state might be, it could be represented symbolically by $|S\rangle$. The angular bracket, $|\rangle$, is called a ket. It is the ket symbol that is important, not so much the symbol inside it. For instance we could just as easily called the state vector $|\sigma\rangle$. The importance of $|S\rangle$ is that it behaves like a complex vector. We shall refer to $|S\rangle$ as the state vector.

The complex conjugate of $|S\rangle$ is $|S\rangle^*$, but this is given a special symbol of its own:

$$|S\rangle^* = \langle S|.$$

The symbol $\langle |$ is a bra. Previously, if a vector $\mathbf{V}$ was normalised we knew that $\mathbf{V}^* \cdot \mathbf{V} = 1$. For a vector $|S\rangle$, this relation becomes

$$\langle S|S\rangle = 1.$$

When a bra and a ket get together, we have a bra-ket, or bracket; hence the terminology. Two different vectors can be orthogonal. In sympathy with $\mathbf{V}_1^* \cdot \mathbf{V}_2 = 0$ we have

$$\langle S_1|S_2\rangle = 0.$$

The normalisation and orthogonality conditions are summarised by

$$\langle S_i|S_j\rangle = \delta_{i.j}.$$

Before proceeding further we should take notice of vectors in Hilbert space. Although they may be complex, we are not allowed to try and find their real and imaginary parts by using a device such as

$$\begin{aligned} \mathrm{Re}|S\rangle &= \tfrac{1}{2}(|S\rangle^* + |S\rangle) \\ &= \tfrac{1}{2}(\langle S| + |S\rangle) \quad \text{(incorrect)} \end{aligned}$$

This is because $\langle S|$ and $|S\rangle$ are said to exist in two different mathematical spaces, dual spaces. It is not possible to add or subtract them as if they belonged to the same space.

The state vector plays a unique part in quantum theory because potentially it provides all the information about a system that we can measure by experiment. We gain knowledge about quantum states by preparing them. For example, if we pass unpolarised light through a piece of polaroid the light which emerges is plane polarised. All the photons which emerge from the polaroid are all polarised in the same sense. They have all emerged in the same state. A good definition of a quantum state has been given by J.M. Jauch:

A state is the result of a series of physical manipulations on the system which constitute the preparation of the state. Two states are identical if the relevant conditions in the preparation of the state are identical.

(J.M. Jauch, *Foundations of Quantum Mechanics*, Addison-Wesley, Reading, Mass., 1968)

We are not usually content with preparing states; we want to measure the properties of systems when they are in a certain state. That is, we need to determine the values of particular observable quantities – the observables of a system. Thus a way has to be found in Dirac's theory to obtain such information from the state vector. The way this is done is to rely on the use of operators. For example, to find the energy of a system it is necessary to elicit the information from the state vector by operating on it with an energy operator. Similarly momentum is found by using a momentum operator, and position by using a position operator. To repeat, in the

Table 6.1. *Some observables and their operator symbols*

Observable	Operator symbol
Energy	$\hat{H}$ (the Hamiltonian)
Linear momentum	$\hat{p}$
Angular momentum	$\hat{\mathbf{L}}$
Position	$\hat{x}$

Subscripts may be added. For example, $\hat{p}_x$ is the operator for linear momentum directed along the x-axis.

construction of the theory, observables are represented by operators; but observable quantities as they are measured are real numbers (scalars). Thus the desired effect of operating on the state vector is to produce a real number. The type of process is similar in kind to that we met in section 6.4, i.e.

$$\text{operator} \times \text{function} = \text{a number} \times \text{the same function.}$$

It will be remembered that functions for which this is true are eigenfunctions of the particular operator, and the number is an eigenvalue. If the function happens to behave like a vector then it is permissible to speak of an eigenvector, or in quantum theory of an eigenstate, Thus,

$$\text{operator} \times \text{eigenvector}$$
$$= \text{eigenvalue} \times \text{eigenvector.}$$

The set of symbols we shall use to represent certain important operators is given in table 6.1.

We can do a lot of useful work without defining the precise form of the operators. For example, suppose $|N\rangle$ is an eigenvector of the Hamiltonian operator, $\hat{H}$. If it is, then when $\hat{H}$ operates on it, the eigenvalue must be the energy of the state, i.e.

$$\hat{H}|N\rangle = E|N\rangle.$$

Now multiply on the left by $|N\rangle^*$, or $\langle N|$.

$$\langle N|\hat{H}|N\rangle = \langle N|E|N\rangle,$$

but E is just a number so

$$\langle N|\hat{H}|N\rangle = E\langle N|N\rangle.$$

If $|N\rangle$ is normalised we finally obtain

$$E = \langle N|\hat{H}|N\rangle.$$

This is our recipe for determining the energy; but remember $|N\rangle$ has to be an eigenvector of $\hat{H}$ for this to work. By a similar procedure we can produce formulae for the eigenvalues of the other operators. For example, if $|M\rangle$ is an eigenvector of the momentum operator $\hat{p}$, we would find the eigenvalue p

$$p = \langle M|\hat{p}|M\rangle.$$

Note the difference in symbol between p and $\hat{p}$, the 'hat', ^, indicates the operator.

You may have noticed that by choosing different symbols for the eigenvectors of $\hat{H}$ and $\hat{p}$ we have implied that different operators need not have the same set of eigenvectors. In fact they often do share eigenvectors in common.

Questions

6.12 Why is it that the state vector cannot itself correspond to an observable?

6.13 Suppose that $\hat{H}$ and $\hat{p}$ do have an eigenstate in common. Call this state $|S\rangle$. Then $H|S\rangle = E|S\rangle$ and $\hat{p}|S\rangle = p|S\rangle$.
   (i) Evaluate $\langle S|\hat{H}\hat{p}|S\rangle$.
   (ii) Evaluate $\langle S|\hat{p}\hat{H}|S\rangle$.
   (iii) We can write

$$\langle S|\hat{H}\hat{p}|S\rangle - \langle S|\hat{p}\hat{H}|S\rangle$$
$$= \langle S|\hat{H}\hat{p} - \hat{p}\hat{H}|S\rangle$$
$$= \langle S|[\hat{H}, \hat{p}]|S\rangle.$$

   Do $\hat{H}$ and $\hat{p}$ commute?

6.6   Expectation values

In section 6.2 we found that we could write a vector $\mathbf{V}$ in terms of a basis set $\mathbf{v}_i$:

$$\mathbf{V} = \sum c_i \mathbf{v}_i.$$

Similarly, let us take a set of vectors $|N_i\rangle$ which are all eigenvectors of the Hamiltonian energy operator. $|N_1\rangle$ will have energy E_1, $|N_2\rangle$ energy E_2 etc. Now take a linear combination of them to form $|S\rangle$

$$|S\rangle = \sum c_i |N_i\rangle.$$

This is a mathematical way of writing down the principle of superposition. This principle says that it is valid to write a state vector as a linear combination of the eigenvectors of a particular operator.

It follows that

$$\hat{H}|S\rangle = \sum c_i \hat{H}|N_i\rangle$$

which, with a little thought gives

$$\hat{H}|S\rangle = \sum c_i E_i |N_i\rangle.$$

Let us take a specific example. Suppose we carry out an experiment that yields two results for the energy of a system. Call these E_1 and E_2. If we measure E_1 we can be sure that the system will have resulted in possessing the state $|N_1\rangle$ which corresponds to this energy. Similarly, when E_2 occurs, the system must have been prepared in the state $|N_2\rangle$. This we know after the experiment. Before it we cannot tell which state the system is in. The principle of superposition says that the state before the measurement will be a linear combination of $|N_1\rangle$ and $|N_2\rangle$. That is,

$$|S\rangle = c_1 |N_1\rangle + c_2 |N_2\rangle.$$

Then,

$$\hat{H}|S\rangle = c_1 E_1 |N_1\rangle + c_2 E_2 |N_2\rangle$$

and

$$\begin{aligned}\langle S|\hat{H}|S\rangle &= (c_1|N_1\rangle + c_2|N_2\rangle)^* \\ &\quad \cdot (c_1 E_1 |N_1\rangle + c_2 E_2 |N_2\rangle) \\ &= (c_1{}^*\langle N_1| + c_2{}^*\langle N_2|) \\ &\quad \cdot (c_1 E_1 |N_1\rangle + c_2 E_2 |N_2\rangle) \\ &= c_1{}^* c_1 E_1 + c_2{}^* c_2 E_2,\end{aligned}$$

because, as usual, we assume $\langle N_i | N_j \rangle = \delta_{i,j}$. This result appears to say that the energy we expect for the result of the experiment is not E_1 or E_2 alone, but a combination of both. We shall write

$$E_{ex} = c_1{}^* c_1 E_1 + c_2{}^* c_2 E_2$$

where E_{ex} is the expectation energy. According to this definition, also

$$E_{ex} = \langle S|\hat{H}|S\rangle.$$

The problem now is to resolve the apparent disparity between the results of the experiment, which are either E_1 or E_2 and the value for the expectation energy, for clearly they are not the same. This is where we turn to Born's probability interpretation. This reminds us that although we know that E_1 or E_2 will occur, in any given single experiment, we cannot predict with certainty which will actually be observed. Bore proposed that the coefficients $c_1{}^* c_1$ and $c_2{}^* c_2$ give the probabilities of energies E_1 and

E_2 being measured. Let us put some numbers to these quantities.

Suppose when the energies were measured we find $E_1 = 100\,\text{J}$, and $E_2 = 120\,\text{J}$. If the experiment were repeated a large number of times, we might measure E_1 40 times and E_2 60 times. On the basis of this (rather limited) number of experiments we would say that the probability of measuring E_1 is 0.4, and of measuring E_2 is 0.6. Now we shall use a result to be found in table 2.1 for the average of a series of measurements:

$$\mu = \sum x_i P(x_i).$$

In this case, x_i is the energy, so

$$\mu = E_1 P(E_1) + E_2 P(E_2)$$

or

$$\begin{aligned}\mu &= P(E_1)E_1 + P(E_2)E_2 \\ &= 0.4 E_1 + 0.6 E_2 \\ &= 112\,\text{J}.\end{aligned}$$

It is very tempting, and we believe correct, to draw a one to one correspondence between the two equations

$$\mu = P(E_1)E_1 + P(E_2)E_2$$

and

$$E_{ex} = c_1{}^* c_1 E_1 + c_2{}^* c_2 E_2$$

so that $c_1{}^* c_1$ corresponds to the probability $P(E_1)$ and $c_2{}^* c_2$ to $P(E_2)$.

We have called μ the average energy while the corresponding quantity from quantum theory is called the expectation value of the energy. They are one and the same thing. Thus the expectation value is 112 J also even though the actual measurements are 100 J or 120 J. The terminology is very widely used, although as we see the name 'expectation value' of the energy is a little misleading. In fact we would not expect to measure the expectation value of the energy. However there is one case when the expectation value would literally be expected to be obtained from experiment. This occurs when the state $|S\rangle$ is itself an eigenvector of the Hamiltonian. Then,

$$\hat{H}|S\rangle = E_S |S\rangle$$

and

$$\langle S|\hat{H}|S\rangle = E_S$$

where E_S is the expectation value for the energy of the state $|S\rangle$.

Questions

6.14 In section 6.6 we wrote $\hat{H}|S\rangle = \sum c_i E_i |N_i\rangle$ with $|S\rangle = \sum c_i |N_i\rangle$. Look back at section 6.6 and try to write down a general equation for $\langle S|\hat{H}|S\rangle$.

6.15 (i) If Born's probability theory is correct, what must be the value of $c_1{}^* c_1 + c_2{}^* c_2$ in the example we considered?

(ii) Using $|S\rangle = \sum c_i |N_i\rangle$, relate the answer you have just given to the demand that $\langle S|S\rangle = 1$.

(iii) Thus explain why there is so much emphasis on normalisation in quantum theory.

6.16 A student discovered a state $|S\rangle$ which was an eigenvector of $\hat{H}$ ($\hat{H}|S\rangle = E|S\rangle$) and was very pleased with her discovery. Shortly afterwards another student claimed to have discovered another state, $|S'\rangle$, of the same system, with exactly the same energy:

$$\hat{H}|S'\rangle = E|S'\rangle.$$

After some thought they discovered that there were in fact an infinite number of eigenstates with the same energy. How could this be? Hint: write $|S'\rangle = e^{i\phi}|S\rangle$.

6.7 Measurement

When a measurement was made of the energy of the simple system we discussed previously we found that the result had to be either E_1 or E_2, but not both. Just prior to making a measurement we could predict with certainty that E_1 or E_2 will occur, but not which: we can only say that it is more probable that E_2 will occur than E_1. The crucial point is that unlike in classical physics where we can apparently predict with certainty the outcome of an experiment once the initial state is known, in quantum theory we cannot foretell the outcome of an experiment with such certainty. What is more, we cannot do anything about the situation. For example, we cannot attempt to modify the apparatus in order to measure E_1 or E_2 alone each and every time. Nor can we propose that there is a single 'perfect' result which can be obtained by correcting for E_1 and E_2 as if they differed from the expectation value because of experimental error. The reason why such an important difference arises between quantum and classical physics is that whereas the classical physicist believes himself capable of allowing for any influence he might have on the system being investigated, the quantum physicist is faced with a situation in which the very act of measurement brings about a change in the system which cannot in principle be controlled. Thus in our example, prior to the measurement of the energy, the system was assumed to be in the state $|S\rangle$. After the measurement we find that we have observed an energy E_1 or E_2. If we observe E_1 then it must be the case that the system will be found in the state whose eigenvalue is E_1, i.e. the new state must be $|N_1\rangle$.

Fig. 6.3. The effect of measurement of the energy on a state composed of two eigenstates of the Hamiltonian.

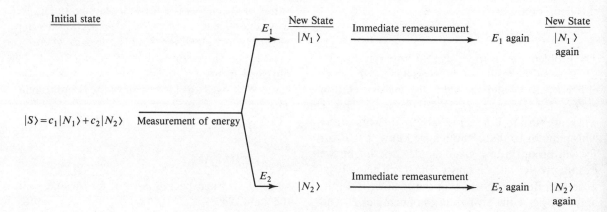

This process can be illustrated as follows:

Just before measurement:

state is $|\psi\rangle = \sum c_i |N_i\rangle$.

Just after measurement:

$\begin{cases} \text{state is } |N_1\rangle \text{ if } E_1 \text{ was the result} \\ \text{state is } |N_2\rangle \text{ if } E_2 \text{ was the result.} \end{cases}$

Thus another strange facet of quantum theory appears. We can give little detailed information about the state of a system which existed just before the measurement. After the measurement we can say with certainty that the system is in state $|N_1\rangle$ if we measure E_1 but we cannot deduce that the state of the system was simply $|N_1\rangle$ before the measurement. The brutal fact is that we just do not know the prior state with certainty. On the more positive side we can see that by making a measurement we have in effect prepared a state. For example, the measurement of energy E_1 effectively prepares the state of the system to be $|N_1\rangle$ after the measurement. Then if immediately after the measurement we make another measurement of the energy, we can predict with certainty that it will again be E_1 because the new state $|N_1\rangle$ is an eigenstate of the Hamiltonian. The situation is represented in fig. 6.3. In a sense then the state of the system 'collapses' from $|S\rangle$ to either $|N_1\rangle$ or $|N_2\rangle$. If we were dealing with wavefunctions, an analogous situation would be where a number of waves interfered to produce a wavepacket. After the measurement only one wavefunction survives. This is known as the collapse of the wavepacket, or wavefunction, of the original system.

6.8   Representations

One of the beauties of Dirac's approach is that we do not always need to know the precise form that the operators corresponding to observables take. However, it is often highly convenient to stipulate the nature of the operators. There are two main ways of doing this. The first, and most popular, is called the Schrödinger representation. The second is the Heisenberg representation. The former represents the position operator, $\hat{x}$, of table 6.1, as equivalent to multiplication by x in our usual Cartesian coordinate system. This is particularly simple and therein lies its popularity. However, in

Table 6.2. *Operators in Schrödinger's representation corresponding to classical terms*

Classical term	General operator form	Operator form in the Schrödinger representation
Position, x	$\hat{x}$	x (same as classical term)
Linear momentum, mv_x or p_x	$\hat{p}_x$	$\dfrac{\hbar}{i}\dfrac{\partial}{\partial x}$
Square of linear momentum, e.g. $p_x{}^2$	$\hat{p}_x{}^2$	$-\hbar^2\dfrac{\partial^2}{\partial x^2}$
Angular momentum, L_z	$\hat{L}_z$	$\dfrac{\hbar}{i}\dfrac{\partial}{\partial\phi}$ or $\dfrac{\hbar}{i}\left(x\dfrac{\partial}{\partial y}-y\dfrac{\partial}{\partial x}\right)$
Square of angular moment, $L_z{}^2$	$\hat{L}_z{}^2$	$-\hbar^2\dfrac{\partial^2}{\partial\phi^2}$ or $-\hbar^2\left(x\dfrac{\partial}{\partial y}-y\dfrac{\partial}{\partial x}\right)\left(x\dfrac{\partial}{\partial y}-y\dfrac{\partial}{\partial x}\right)$
Kinetic energy, $\dfrac{p^2}{2m}$	$\hat{H}$	$-\dfrac{\hbar^2}{2m}\nabla^2$

Schrödinger's system, linear and angular momentum are represented by differential operators. For example, $\hat{p}_x$ becomes $(\hbar/i)(\partial/\partial x)$. The form of the common operators in Schrödinger's representation are collected together in table 6.2. Use of these operators also gives us a simple way of converting classical formulae into their quantum counterparts.

If we consider a freely-moving particle, with momentum $p_x = mv_x$, then its energy is $p_x{}^2/2m$. This is the classical recipe. All we have to do is measure the momentum and mass, perform an easy calculation and we have the value for the energy. In quantum theory we know that the value of the momentum should be obtained as an eigenvalue of the operator $\hat{p}_x$. Similarly, the energy will be obtained as an eigenvalue of $\hat{p}_x{}^2/2m$. From table 6.2 we find

$$\frac{\hat{p}_x{}^2}{2m} \equiv -\frac{\hbar^2}{2m}\frac{\partial^2}{\partial x^2}.$$

We can quickly recognise this as the type of term we found in the introduction to Schrödinger's equation of chapter 2. There is another feature of

Table 6.3. *The correspondence between expressions in the Dirac and Schrödinger methods*

Expectation Value of	Dirac	Schrödinger
Position, x	$\langle S\|\hat{x}\|S\rangle$	$\int \psi^* x \psi \, dx$
Linear momentum, p_x	$\langle S\|\hat{p}_x\|S\rangle$	$\int \psi^* \hat{p}_x \psi \, dx$
		$= \dfrac{\hbar}{i}\int \psi^* \dfrac{\partial}{\partial x}\psi \, dx$
Angular momentum, L_z	$\langle S\|\hat{L}_z\|S\rangle$	$\int \psi^* \hat{L}_z \psi \, dv$
		$= \dfrac{\hbar}{i}\int \psi^* \dfrac{\partial}{\partial \varphi}\psi \, dv$
Kinetic energy, $\dfrac{p^2}{2m}$	$\langle S\|\hat{H}\|S\rangle$	$\int \psi^* \hat{H}\psi \, dv$
	or $\dfrac{1}{2m}\langle S\|\hat{p}^2\|S\rangle$	or $\dfrac{1}{2m}\int \psi^* \hat{p}^2 \psi \, dv$
		$= -\dfrac{\hbar^2}{2m}\int \psi^* \nabla^2 \psi \, dv$

Note also the equivalence of
(i) the normalisation conditions $\langle S_i|S_i\rangle = 1$ and $\int \psi_i^*\psi_i \, dv = 1$
(ii) the orthogonalisation conditions

$$\langle S_i|S_j\rangle = 0 \text{ and } \int \psi_i^*\psi_j \, dv = 0 \text{ for } i \neq j$$

(iii) formulae for transition dipole moments such as

$$\langle S_2|\hat{\mathbf{d}}|S_1\rangle \text{ and } \int \psi_2^* \hat{\mathbf{d}}\psi_1 \, dv \text{ where } \hat{\mathbf{d}} = -e\hat{\mathbf{r}}.$$

Schrödinger's representation which we shall not attempt to prove. This is that his choice of position operator results in a special type of equivalence between his and Dirac's methods of solving problems. The simplest way of seeing this is to look at table 6.3. Instead of dealing with complex vectors we use complex wavefunctions, in place of scalar products we have integration.

In passing we should at least consider the Heisenberg representation. In this representation the momentum operator is *not* a differential operator but the position operator *is* a differential operator. For instance in the Heisenberg repre-

sentation $\hat{x} \equiv (\hbar/i)(\partial/\partial p_H)$ where p_H is the momentum in the same representation. It should be obvious why the Heisenberg representation is not too popular! However, the fact that such a representation is possible illustrates the point that the form of the operators in quantum theory is not unique but open to choice. Just as it took many years for classical physics to develop as a science and find useful definitions of quantities such as energy and momentum by putting them as products of mass and powers of velocity, so it has been the task of quantum theory to find useful and self-consistent choices for the operators $\hat{H}$, $\hat{p}$ and $\hat{x}$ (amongst others). Although the formalism of quantum theory appears more abstract than that of classical physics, relying as it does so heavily upon the use of differential operators, it should be remembered that the final test of a physical theory is whether it agrees with the results of experiments. By and large quantum theory passes the test; classical theory does not.

Questions

6.17 The problem is to calculate the expectation value of the position of a particle in a one-dimensional box. Assume that the particle is in the ground state. We can do this as follows:
(i) Write down the expression for x_{ex} in the Schrödinger representation.
(ii) Put in the explicit form for the wavefunction, and evaluate the integral (see section 2.3). Comment on your answer.

6.18 Repeat the method of the last question to calculate the expectation value of the momentum. Again comment on the results.

6.19 In atomic units, a hydrogen 1s orbital is $(1/\sqrt{\pi})e^{-r}$. For this orbital we can neglect the part of the Hamiltonian that depends on θ and ϕ (see section 3.1). The kinetic energy operator, $\hat{\mathscr{T}}$, becomes

$$\hat{\mathscr{T}} = -\frac{1}{2}\frac{1}{r^2}\frac{\partial}{\partial r^2}\left(r^2\frac{\partial}{\partial r}\right).$$

Similarly, the potential energy operator is $-1/r^2$.
(i) Calculate the expectation value of $\hat{\mathscr{T}}$.
(ii) Calculate the expectation value of $\hat{V}$.

(iii) Is the virial theorem obeyed (see section 1.5)?

Hints: Integration must be over $d\upsilon = r^2\,dr\sin\theta\,d\theta\,d\phi$. Use the standard integrals in appendix A.

6.9   Operators and the uncertainty relations

The operators that appear in quantum theory have to obey certain rules if they are to represent observables. The main rules are collected here in the form of five theorems.

Theorem 1

An operator corresponding to an observable must be linear.

For example, if $\hat{A}$ is the operator and $|V_1\rangle$, $|V_2\rangle$ are two of its eigenvectors we must have

$$\hat{A}(|V_1\rangle + |V_2\rangle) = \hat{A}|V_1\rangle + \hat{A}|V_2\rangle.$$

This bears comparison with the work in section 6.6.

Theorem 2

An operator corresponding to an observable must be Hermitian.

To see what this means, let our operator $\hat{A}$ operate on $|V_1\rangle$ to give $\hat{A}|V_1\rangle$. Now act on the left hand side with the bra, $\langle V_2|$, of another vector. This gives $\langle V_2|\hat{A}|V_1\rangle$. If $\hat{A}$ is Hermitian it has the property that $\langle V_2|\hat{A}|V_1\rangle = \langle V_1|A|V_2\rangle^*$. (The same relations holds if $|V_1\rangle = |V_2\rangle$.) The point of this shows in:

Theorem 3

The eigenvalues of an Hermitian operator are real.

Suppose $\hat{A}$ has a eigenvector $|V_1\rangle$ which yields the eigenvalue a_1, i.e.

$$\hat{A}|V_1\rangle = a_1|V_1\rangle.$$

Then

$$\langle V_1|\hat{A}|V_1\rangle = a_1\langle V_1|V_1\rangle$$

or

$$\langle V_1|\hat{A}|V_1\rangle = a_1. \tag{B}$$

Now take the complex conjugate of both sides to give,

$$\langle V_1|\hat{A}|V_1\rangle^* = a_1^*. \tag{C}$$

But if $\hat{A}$ is Hermitian,

$$\langle V_1|\hat{A}|V_1\rangle = \langle V_1|\hat{A}|V_1\rangle^*.$$

Thus, equation (C) gives

$$\langle V_1|A|V_1\rangle = a_1^*$$

and comparing this with equation (B) shows that

$$a_1 = a_1^*.$$

The only type of number which equals its own complex conjugate is a real number. Therefore a_1 must be real. This proves the theorem. The reason why it is important is that the values of observables must be real numbers. In this case a_1 could be the value of an observable; if a_1 were complex it could not.

We have shown that an operator corresponding to an observable must be Hermitian.

Theorem 4

If an Hermitian operator has a set of eigenvectors, each with different eigenvalues, then the eigenvectors must be orthogonal.

For instance, suppose $|V_1\rangle$ and $|V_2\rangle$ are two eigenvectors of $\hat{A}$ with different eigenvalues, a_1, a_2.

$$\hat{A}|V_1\rangle = a_1|V_1\rangle \text{ and } \hat{A}|V_2\rangle = a_2|V_2\rangle$$

the theorem says that $\langle V_2|V_1\rangle = 0$. We can see this in operation in section 6.5. The theorem is proved in M.6.1.

Theorem 5

If there exists a set of orthogonal eigenvectors which are eigenvectors of two operators $\hat{A}$, $\hat{B}$ say then $\hat{A}$ and $\hat{B}$ must commute, i.e. $[\hat{A},\hat{B}] = [\hat{A}\hat{B} - \hat{B}\hat{A}] = 0$.

This too is proved in M.6.1. There is a converse to this theorem: if two operators commute then they have a set of eigenvectors in common which are eigenvectors of both operators.

In the questions in section 6.4 we found that some combinations of operators did not commute. As a further illustration let us take $[\hat{p}_x, \hat{x}]$. In the Schrödinger representation this becomes $[(\hbar/i)\partial/\partial x, x]$. In question 6.11, we showed, that

$$\left[\frac{\hbar}{i}\frac{\partial}{\partial x}, x\right] = \frac{\hbar}{i}.$$

Similarly we can show that

$$[\hat{p}_x{}^2, x] = (2\hbar/\text{i})\hat{p}_x.$$

and

$$[\hat{H}, x] = (\hbar/m\text{i})p_x.$$

Notice that each of these three results is of the form $[\hat{A}, \hat{B}] = \text{i}\hat{C}$ where $\hat{A}$, $\hat{B}$ and $\hat{C}$ are all operators. (In the first case, $\hat{C} \equiv -\hbar$. This is the simple operator which says multiply by the value $-\hbar$.) Combinations of non-commuting operators which behave like this are the subject of Heisenberg's uncertainty relations. If ΔA is the standard deviation of the results which we obtain by measuring the observable corresponding to $\hat{A}$, and if ΔB is the corresponding standard deviation for $\hat{B}$, then

$$\Delta A \Delta B \geq \tfrac{1}{2}|C_{\text{ex}}|.$$

Here $|C_{\text{ex}}|$ is the magnitude of the expectation value of the operator $\hat{C}$ appearing in $[\hat{A}, \hat{B}] = \text{i}\hat{C}$. (Note that $|C_{\text{ex}}|^2 = C_{\text{ex}}^* C_{\text{ex}}$.) Using our previous examples for $\hat{p}_x$ and x,

$$\Delta p_x \Delta x \geq \tfrac{1}{2}|-\hbar|$$

or

$$\Delta p_x \Delta x \geq \tfrac{1}{2}\hbar.$$

This is just one of many relations between variables, such as momentum and position, which are said to be conjugate to one another. The one above was first derived by Heisenberg. As a mathematical relationship it may appear unexceptionable, but its interpretation has remained a matter of debate ever since Heisenberg first published it in 1927.

6.10 Heisenberg's uncertainty principle

A particular case of the uncertainty relation

$$\Delta A \Delta B \geq \tfrac{1}{2}|C_{\text{ex}}|$$

is when $\hat{A}$ and $\hat{B}$ are the momentum and position operators $\hat{p}_x$ and $\hat{x}$. Then, as we have seen,

$$\Delta p_x \Delta x \geq \tfrac{1}{2}\hbar. \tag{D}$$

Of primary importance is the fact that an uncertainty relation like this is an unavoidable result of the way quantum theory works. Indeed, Heisenberg was inclined to put the uncertainty relation as the basic equation of quantum theory, but this probably exaggerates the position. Certainly in the classical physics of particles no such relation between position and momentum occurs. That is, in classical physics position and momentum always commute, e.g. $mv_x x = xmv_x$. Thus in quantum theory the uncertainty relation deserves special attention and some degree of explanation in physical terms.

At this point we should bear in mind that there are at present a number of conflicting interpretations of the uncertainty relations. The one which currently has the most support corresponds to the manner in which Heisenberg originally interpreted the momentum–position uncertainty relation. Heisenberg's interpretation is often the only one discussed in text books and is called Heisenberg's uncertainty principle. Very often equation (D) is also known by this name, but at the risk of being pedantic it should be realised that it represents a mathematical relationship between noncommuting observables in quantum theory. The term 'uncertainty principle' should really be reserved for Heisenberg's interpretation of the relationship. There is no argument about the mathematical validity of uncertainty relations; but there is disagreement about the validity of Heisenberg's interpretation.

Briefly stated, Heisenberg's uncertainty principle is:

The momentum and position of a particle cannot both be measured at the same time to an arbitrary degree of accuracy.

The momentum and position must refer to the same coordinate axis, e.g. p_x and x, p_y and y, or p_z and z; members of pairs like this are said to be conjugate to each other. The principle claims that it may be perfectly feasible to measure the momentum, of, say, an electron with as great an accuracy as possible, so that $\Delta p_x \to 0$, but if this is done a simultaneous measurement of the position of the electron will become increasingly less accurate, and $\Delta x \to \infty$. In a particular case suppose Δp_x is 10^{-20} kg m s^{-1}, then

$$\Delta x \geq 10^{20}\, \hbar/2.$$

Here the smallest value of Δx that could be obtained from a simultaneous measurement of the position would be $10^{20}\hbar/2 \approx 10^{-14}$ m. Note that the uncertainty principle does not say that this minimum will necessarily be obtained – the inaccuracy, or uncertainty, Δx may be greater than 10^{-14} m.

Questions

6.20 Throughout our work we have ignored the possibility of states, or wavefunctions, changing with time. We have concentrated on the properties of stationary states. When time becomes important, e.g. during a transition between states, Schrödinger's equation takes the form

$$\hat{H}_t \psi(t) = i\hbar \frac{\partial \psi(t)}{\partial t}.$$

Then

$$E\psi(t) = i\hbar \frac{\partial \psi(t)}{\partial t}$$

allows us to calculate the energy. The strange thing is that, in spite of this, there is no quantum-mechanical operator corresponding to time. This implies that time has a dubious place as an observable. In spite of this an energy–time uncertainty relation is widely used.

 (i) Using $\hat{H}_t = i\hbar(\partial/\partial t)$ and taking 't' as the 'time operator' show that $[\hat{H}_t, t]$ is of the form $[\hat{A}, \hat{B}] = i\hat{C}$.
 (ii) Hence write down the energy–time uncertainty relation. (Use the symbol ΔE instead of ΔH_t.)
 (iii) Show that an alternative form is

$$\Delta f \Delta t \geq \frac{1}{4\pi}.$$

 (iv) Using this relation calculate the least value of Δf for states having a lifetime of 3×10^{-8} s and 10 s.
 (v) What is the ratio of Δf in these cases. This gives an estimate of the ratio of the widths of spectral lines in atomic and nuclear magnetic resonance spectroscopy.

6.11 Zero point energy

The lowest energy which a simple harmonic oscillator can have is $\frac{1}{2}hf$. Even at absolute zero the oscillator will have this residual amount of energy. The uncertainty principle has been applied in this instance to show that the existence of the zero point energy is in accord with the principle. If there were no zero point energy then the mass (or masses) comprising the oscillator would come to rest at absolute zero. If this were the case they would have a definite position which could be measured and also a definite momentum, namely zero. Thus without a point energy both Δp_x and Δx could be simultaneously measured with any desirable accuracy and the uncertainty principle would be violated. Where there is the zero point energy there remains a spread of values of momentum and position and the uncertainty principle remains intact.

6.12 Summary

This chapter has been one where we have generated many results which are of a very general nature. We found that we could obtain information from Dirac's eigenvectors by using operators to act on them. If operators are to give us the values of observables they have to have eigenvalues which are real numbers. This restriction, and others, meant that operators have to satisfy a number of conditions which we summarised in five theorems.

We found that Dirac's theory could be given two major representations, the main one being due to Schrödinger. In Schrödinger's representation the operators take on particular forms which we can use to adapt classical formulae to serve in quantum theory.

Just as in the case of Schrödinger's wavefunctions, which mainly concerned us in earlier chapters, Dirac's method can only be given a satisfactory interpretation by adopting Born's probability model. Quantum theory is deeply involved in statistics.

M.6.1

To prove theorem 4

With $\hat{A}|V_1\rangle = a_1|V_1\rangle$ we obtain

$$\langle V_2|\hat{A}|V_1\rangle = a_1\langle V_2|V_1\rangle.$$

Similarly, if $\hat{A}|V_2\rangle = a_2|V_2\rangle$

$$\langle V_1|\hat{A}|V_2\rangle = a_2\langle V_1|V_2\rangle.$$

But $\hat{A}$ is an Hermitian operator so

$$\langle V_2|\hat{A}|V_1\rangle^* = a_1^*\langle V_2|V_1\rangle^*$$

means

$$\langle V_1|\hat{A}|V_2\rangle = a_1\langle V_1|V_2\rangle.$$

Thus we have

$$a_2\langle V_1|V_2\rangle = a_1\langle V_1|V_2\rangle$$

or

$$(a_1 - a_2)\langle V_1|V_2\rangle = 0.$$

We have assumed that $a_1 \neq a_2$ so it must be that $\langle V_1|V_2\rangle = 0$. We have proved that the eigenvectors must be orthogonal.

To prove theorem 5

Suppose $|V_1\rangle$ is an eigenvector of $\hat{A}$ and $\hat{B}$. Then,

$$\hat{A}|V_1\rangle = a_1|V_1\rangle$$

and

$$\hat{B}\hat{A}|V_1\rangle = a_1\hat{B}|V_1\rangle$$
$$= a_1 b_1|V_1\rangle.$$

Similarly we have

$$\hat{B}|V_1\rangle = b_1|V_1\rangle$$

and

$$\hat{A}\hat{B}|V_1\rangle = b_1\hat{A}|V_1\rangle$$
$$= b_1 a_1|V_1\rangle$$

but $a_1 b_1 = b_1 a_1$ because real numbers always commute.

This means that

$$\hat{A}\hat{B}|V_1\rangle - \hat{B}\hat{A}|V_1\rangle = (a_1 b_1 - b_1 a_1)|V_1\rangle.$$

As a consequence, $\hat{A}\hat{B} - \hat{B}\hat{A} = 0$. This proves that $\hat{A}$ and $\hat{B}$ commute.

7

Advanced methods of quantum chemistry

7.1 Introduction

In this chapter we shall find quantum theory applied to chemistry in some powerful ways. The role of angular momentum is so vital to our understanding of the properties of atoms and molecules that we shall delve deeper into its treatment in operator theory. This will give us some powerful results which we can use to provide mathematical substance to several effects, such as n.m.r., that we met in earlier chapters.

The Zeeman effect has been of great use in helping spectroscopists to unravel the mysteries of atomic structure. We shall discover how to calculate the influence that a magnetic field has on an atom. Having shown that a magnetic field will perturb the energy of an atom or molecule we shall go on to develop perturbation theory, and apply it in some simple cases.

Finally, as promised in chapter 2, we shall prove the presence of zero point energy in a simple harmonic oscillator. This represents something of a *tour de force* of operator theory.

7.2 Angular momentum

One of the differences between Bohr's and Schrödinger's work on the hydrogen atom was that Bohr claimed that the angular momentum of the electron was given by a formula of the type

$$L = l\hbar$$

while Schrödinger's result was that

$$L = \sqrt{[l(l + 1)]}\hbar.$$

We did not prove Schrödinger's formula to be the correct one, but we know that it gives better agreement with experiment than does Bohr's. Here we shall prove that L is given by $\sqrt{[l(l + 1)]}\hbar$.

First, we can change the problem a little because, in Dirac's method, we have to show that the eigenvalues of the operator $\hat{\mathbf{L}}^2$ are $l(l + 1)\hbar^2$. Our method takes us on a detour through the properties of operators. We shall break the task into four stages.

Stage 1: Some new operators

Two operators we shall write as $\hat{L}_+$ and $\hat{L}_-$ are very useful. They are called ladder operators. $\hat{L}_+$ is a step-up, and $\hat{L}_-$ a step-down operator. We shall see why they are given these names shortly. They are defined by

$$\hat{L}_+ = \hat{L}_x + i\hat{L}_y.$$
$$\hat{L}_- = \hat{L}_x - i\hat{L}_y.$$

From these definitions we find that

$$\hat{L}_+\hat{L}_- = \hat{L}_x{}^2 + L_y{}^2 + i(\hat{L}_y\hat{L}_x - \hat{L}_x\hat{L}_y).$$

Therefore

$$\hat{L}_+\hat{L}_- = \hat{L}_x{}^2 + \hat{L}_y{}^2 + \hbar\hat{L}_z$$
$$= \hat{\mathbf{L}}^2 - \hat{L}_z{}^2 + \hbar\hat{L}_z$$

because $[\hat{L}_y, \hat{L}_x] = (\hbar/i)\hat{L}_z$ (see section 6.4) and $\hat{\mathbf{L}}^2 = \hat{L}_x{}^2 + \hat{L}_y{}^2 + \hat{L}_z{}^2$. Also in section 6.4 we find that $\hat{\mathbf{L}}^2$ and $\hat{L}_z$ commute. From theorem 5, section 6.9 it follows that they can have a set of eigenvectors in common. We shall call a general member of this set $|a, b\rangle$ where a gives the eigenvalues of $\hat{\mathbf{L}}^2$ and b gives the eigenvalues of $\hat{L}_z$:

$$\hat{\mathbf{L}}^2|a, b\rangle = a|a, b\rangle$$
$$\hat{L}_z|a, b\rangle = b|a, b\rangle.$$

In M.7.1 we find that

$$[\hat{L}_z, \hat{L}_+] = \hbar\hat{L}_+.$$

We shall use this in stage 2.

Stage 2: Ladder operators at work

Expanding $[\hat{L}_z, \hat{L}_+]$ means that

$$\hat{L}_z\hat{L}_+ - \hat{L}_+\hat{L}_z = \hbar\hat{L}_+.$$

If we operate on $|a, b\rangle$,

$$\hat{L}_z \hat{L}_+ |a, b\rangle - \hat{L}_+ \hat{L}_z |a, b\rangle = \hbar \hat{L}_+ |a, b\rangle.$$

Thus,

$$\hat{L}_z \hat{L}_+ |a, b\rangle - \hat{L}_+ b |a, b\rangle = \hbar \hat{L}_+ |a, b\rangle.$$

Rearranging,

$$\hat{L}_z \hat{L}_+ |a, b\rangle = (b + \hbar) \hat{L}_+ |a, b\rangle.$$

If we write this in a slightly different way,

$$\hat{L}_z (\hat{L}_+ |a, b\rangle) = (b + \hbar)(\hat{L}_+ |a, b\rangle). \tag{A}$$

In words, this equation says

$$\hat{L}_z \text{ operating on } \hat{L}_+ |a, b\rangle$$
$$= \text{a number} \times \hat{L}_+ |a, b\rangle.$$

We can recognise this as a typical eigenvalue equation. The combination $\hat{L}_+ |a, b\rangle$ acts as an eigenvector of $\hat{L}_z$ and its eigenvalue is $(b + \hbar)$. Now we ask the question, 'what must be the form of the vector $|a, b\rangle$ which gives an eigenvalue of $\hat{L}_z$ equal to $(b + \hbar)$?'. It must be $|a, (b + \hbar)\rangle$ because then

$$\hat{L}_z |a, (b + \hbar)\rangle = (b + \hbar)|a, (b + \hbar)\rangle.$$

If we compare this with equation (A) above, we see that $\hat{L}_+$ operating on $|a, b\rangle$ must change it to $|a, (b + \hbar)\rangle$ as well as giving some eigenvalue as well. That is,

$$\hat{L}_+ |a, b\rangle = \text{an eigenvalue} \times |a, (b + \hbar)\rangle.$$

This eigenvalue we shall write N_+ so

$$\hat{L}_+ |a, b\rangle = N_+ |a, (b + \hbar)\rangle.$$

The reason why $\hat{L}_+$ is called a step-up operator is apparent. Every time it acts on a vector $|a, b\rangle$ it steps up the value of b by $\hbar$. For example,

$$\hat{L}_+ |a, b\rangle = N_+ |a, (b + \hbar)\rangle,$$

$$\hat{L}_+ |a, (b + \hbar)\rangle = N_+ |a, (b + 2\hbar)\rangle$$

and so on. By an exactly analogous argument we can show that, as we might guess, $\hat{L}_-$ does the following:

$$\hat{L}_- |a, b\rangle = N_- |a, (b - \hbar)\rangle.$$

It steps down the value of b by $\hbar$ each time.

Every ladder has a top and a bottom which we cannot go beyond. This is true of the ladder operators as well.

Stage 3: Over the top, or off the bottom?

Using our earlier result

$$L_+ L_- |a, b\rangle = (\hat{L}^2 - \hat{L}_z^2 + \hbar \hat{L}_z)|a, b\rangle$$
$$= (a - b^2 + \hbar b)|a, b\rangle.$$

Therefore, $\langle a, b | L_+ L_- | a, b \rangle = a - b^2 + \hbar b$ if $|a, b\rangle$ is normalised. The left hand side of this equation has the form $\langle S | S \rangle$ if we associate $|S\rangle$ with $(\hat{L}_- |a, b\rangle)$. We know that the product of a bra and its corresponding ket can never be negative. This allows us to say that

$$a - b^2 + \hbar b \geqslant 0.$$

If we follow through an entirely similar argument we find that $\hat{L}_- \hat{L}_+ |a, b\rangle = a - b^2 - \hbar b$. Therefore,

$$a - b^2 - \hbar b \geqslant 0$$

as well. Adding the two inequalities and dividing by 2

$$a - b^2 \geqslant 0$$

or

$$a \geqslant b^2.$$

This shows us that if, for example $a = 100$ then b could never be greater than $+ 10$. Similarly, if the minimum value of a was 9, b could never be less than $- 3$. In any particular case b will therefore have a maximum value, b_{max} say, and a minimum value, b_{min}. This has the effect of giving us the results

$$\hat{L}_+ |a, b_{max}\rangle = 0$$

and

$$\hat{L}_- |a, b_{min}\rangle = 0.$$

In this way we have found the top and bottom of the eigenvalue ladder. These last two equations say that we cannot go off the top or off the bottom.

Stage 4: The final result

Using our recent findings, we have

$$\langle a, b_{min} | \hat{L}_+ \hat{L}_- | a, b_{min} \rangle = a - b_{min}^2 - \hbar b_{min}$$
$$\langle a, b_{max} | \hat{L}_- \hat{L}_+ | a, b_{max} \rangle = a - b_{max}^2 + \hbar b_{max}$$

on the one hand, but also

$$\langle a, b_{min} | \hat{L}_+ \hat{L}_- | a, b_{min} \rangle = 0$$

and

$$\langle a, b_{max} | \hat{L}_- \hat{L}_+ | a, b_{max} \rangle = 0$$

because of the action of $\hat{L}_-$ on $|a, b_{min}\rangle$ and $\hat{L}_+$ on $|a, b_{max}\rangle$.

Therefore

$$a - b_{min}^2 - \hbar b_{min} = 0$$

and

$$a - b_{max}^2 + \hbar b_{max} = 0. \right\} \qquad \text{(B)}$$

These equations can only be consistent if

$$b_{min} = - b_{max}.$$

Now imagine acting on $|a, b_{min}\rangle$ with $\hat{L}_+$. We could go from b_{min} up to b_{max} in steps of $\hbar$. Each time $\hat{L}_+$ did this we would get a new eigenvalue of $\hat{L}_z$. In fact we would obtain the series

$$b_{min}, b_{min} + \hbar, \ldots, -\hbar, 0, \hbar, \ldots, b_{max} - \hbar, b_{max}.$$

If we write b_{max} as $l\hbar$, then b_{min} will be $-l\hbar$ and we obtain the series

$$-l\hbar, (-l+1)\hbar, \ldots, -\hbar, 0,$$
$$\hbar, \ldots, (l-1)\hbar, l\hbar.$$

There are $2l + 1$ of these values. From either of the equations (B), we have

$$a = (l\hbar)^2 + l\hbar^2$$

or

$$a = l(l+1)\hbar^2.$$

We shall label the $2l + 1$ values of b as a whole number of multiples of $\hbar$. That is, we shall put $b \equiv m_l\hbar$ where m_l, like l, is an integer. Because of these changes in notation, it makes more sense to label the eigenvectors of $\hat{L}^2$ and $\hat{L}_z$ as $|l, m_l\rangle$ rather than $|a, b\rangle$. We can summarise our results as follows:

$$\hat{L}^2|l, m_l\rangle = l(l+1)\hbar^2|l, m_l\rangle$$
$$\hat{L}_z|l, m_l\rangle = m_l\hbar|l, m_l\rangle$$
$$\hat{L}_+|l, m_l\rangle = N_+|l, (m_l + 1)\rangle$$
$$\hat{L}_-|l, m_l\rangle = N_-|l, (m_l - 1)\rangle.$$

We could use exactly the same arguments to prove that a similar set of relationships hold for the spin angular momentum operators and spin vectors:

$$\hat{S}^2|s, m_s\rangle = s(s+1)\hbar^2|s, m_s\rangle$$
$$\hat{S}_z|s, m_s\rangle = m_s\hbar|s, m_s\rangle$$
$$\hat{S}_+|s, m_s\rangle = N_+|s, (m_s + 1)\rangle$$
$$\hat{S}_-|s, m_s\rangle = N_-|s, (m_s - 1)\rangle.$$

If

$$\hat{J} = \hat{L} + \hat{S}$$
$$\hat{J}^2|j, m_j\rangle = j(j+1)\hbar^2|j, m_j\rangle$$

$$\hat{J}_z|j, m_j\rangle = m_j\hbar|j, m_j\rangle.$$
$$\hat{J}_+|j, m_j\rangle = N_+|j, (m_j + 1)\rangle$$
$$\hat{J}_-|j, m_j\rangle = N_-|j, (m_j - 1)\rangle.$$

In the case of spin vectors $|s, m_s\rangle$ we already know that, for an electron $s = \frac{1}{2}$ while $m_s = \frac{1}{2}$ or $-\frac{1}{2}$ are the only possible combinations. Just as we found in chapter 3 it is convenient to use a different label for spin states. Often it is neater to put

$$|\tfrac{1}{2}, \tfrac{1}{2}\rangle \equiv |\alpha\rangle$$

and

$$|\tfrac{1}{2}, -\tfrac{1}{2}\rangle \equiv |\beta\rangle.$$

Much of our work here has been abstract. As a particular example, which has application in spectroscopy, let us see how the theory can be applied to the plane rigid rotator problem.

If we ignore vibrations, a diatomic molecule can be considered as a plane rigid rotator (fig. 7.1). If it rotates in the xy-plane, its energy according to classical physics is $L_z^2/2I$ with I the moment of inertia. We can quickly convert this to form the Hamiltonian in quantum theory:

$$\hat{H} = \frac{\hat{L}_z^2}{2I}.$$

Writing its eigenvectors as $|l, m\rangle$ where, for convenience we omit the subscript on m, we have

$$\hat{L}_z^2|l, m\rangle = m^2\hbar^2|l, m\rangle.$$

Fig. 7.1. A diatomic molecule, rotating in the xy-plane, can serve as a plane rigid rotator.

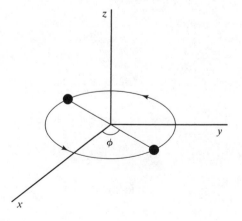

Therefore,

$$\hat{H}|l,m\rangle = \frac{m^2\hbar^2}{2I}\Big|l,m\rangle.$$

But

$$\hat{H}|l,m\rangle = E_m|l,m\rangle$$

so

$$E_m = \frac{m^2\hbar^2}{2I}.$$

This is the formula we used in section 5.5. Again, notice the degeneracy of states with $+m$ and $-m$.

To derive the selection rules for electric dipole transitions of a rigid rotator let us change to the Schrödinger representation. Then,

$$\hat{L}_z \equiv \frac{\hbar}{i}\frac{\partial}{\partial \phi}.$$

It is straightforward to check that a suitable set of eigenfunctions is

$$\psi_m = \frac{1}{\sqrt{(2\pi)}}e^{im\phi}.$$

The transition dipole moment for a transition between $\psi_1 = [1/\sqrt{(2\pi)}]e^{im_1\phi}$ and $\psi_2 = [1/\sqrt{(2\pi)}]e^{im_2\phi}$ is

$$\int_0^{2\pi} \psi_2^*\,\hat{\mathbf{d}}\psi_1\,d\phi = \frac{1}{2\pi}\int_0^{2\pi} e^{-im_2\phi}\,\hat{\mathbf{d}}e^{im_1\phi}d\phi.$$

If we take the x-component of $\hat{\mathbf{d}}$,

$$\hat{d}_x = -ex = -er\cos\phi.$$

Now concentrate on the corresponding integral:

$$-\frac{e}{2\pi}\int_0^{2\pi} e^{-im_2\phi}r\cos\phi\, e^{im_1\phi}\,d\phi$$

$$= -\frac{er}{2\pi}\int_0^{2\pi} e^{i(m_1-m_2)\phi}\cos\phi\,d\phi$$

$$= -\frac{er}{4\pi}\int_0^{2\pi} e^{i(m_1-m_2)\phi}(e^{i\phi}+e^{-i\phi})d\phi$$

$$= -\frac{er}{4\pi}\int_0^{2\pi} [e^{i(m_1-m_2+1)\phi}+e^{i(m_1-m_2-1)\phi}]d\phi$$

But from question 6.2, we know that

$$\int_0^{2\pi} e^{in\phi}d\phi = 0 \qquad \text{unless } n=0.$$

Thus for the integral to be non-zero, either.

$$m_1-m_2+1=0 \qquad \text{or} \qquad m_1-m_2-1=0$$

i.e.

$$m_1-m_2=-1 \qquad \text{or} \qquad m_1-m_2=+1.$$

This proves the selection rule $\Delta m = \pm 1$.

Questions

7.1 What is the result of

(i) $\hat{S}_+|\alpha\rangle$; (ii) $\hat{S}_+|\beta\rangle$; (iii) $\hat{S}_-|\alpha\rangle$; (iv) $\hat{S}_-|\beta\rangle$?

7.2 An a orbital has $l=0$ and $m=0$ so it could be represented by $|0,0\rangle$. Similarly a set of p orbitals is $|1,1\rangle$, $|1,0\rangle$, $|1,-1\rangle$.

(i) Using the notation

$$|x\rangle = \frac{1}{\sqrt{2}}(|1,1\rangle + |1,-1\rangle)$$

$$|y\rangle = \frac{1}{i\sqrt{2}}(|1,1\rangle - |1,-1\rangle)$$

$$|z\rangle = |1,0\rangle$$

show that whereas $|x\rangle$ and $|y\rangle$ are not eigenstates of $\hat{L}_z$, the combinations $(1/\sqrt{2})(|x\rangle + i|y\rangle)$ and $(1/\sqrt{2})(|x\rangle - i|y\rangle)$ are such eigenstates.

(ii) What is the result of $\hat{L}_+$ and $\hat{L}_-$ acting on these combinations of $|x\rangle$ and $|y\rangle$?

7.3 For a three-dimensional rigid rotator, $\hat{H} = \hat{L}^2/2I$.

(i) What are the values of the energies, E_l?

(ii) Compare your formula for the one commonly used in spectroscopy where J is used instead of L, i.e. $E_J = BJ(J+1)$. What value does the rotational constant, B, take?

(iii) Which component of the transition dipole moment must be non-zero for the selection rule $\Delta m = 0$?

7.3 The Zeeman effect

When a magnetic field acts on an atom or molecule it can interact with the magnetic moments of the electrons and nuclei. One of the most important effects of such an interaction involving the electrons is the Zeeman effect. The magnetic moment operator of an electron owing to its orbital motion is $-(e/2m_e)\hat{\mathbf{L}}$; owing to its spin it is

$-(e/m_e)\hat{\mathbf{S}}$ where $\hat{\mathbf{L}}$ and $\hat{\mathbf{S}}$ are the orbital and spin operators. We can combine these together to give the operator

$$\hat{\mathbf{m}} = -\frac{e}{2m_e}(\hat{\mathbf{L}} + 2\hat{\mathbf{S}}).$$

The Hamiltonian for the interaction with a magnetic field, **B**, is

$$\hat{H}_{\text{mag}} = -\hat{\mathbf{m}} \cdot \mathbf{B}$$

or

$$\hat{H}_{\text{mag}} = \frac{e}{2m_e}(\hat{\mathbf{L}} + 2\hat{\mathbf{S}}) \cdot \mathbf{B}.$$

This Hamiltonian is fine provided the orbital and spin moments behave largely independently of one another. In chapter 5 we found that this is not often the case. In Russell–Saunders coupling we had $\mathbf{J} = \mathbf{L} + \mathbf{S}$, or in terms of operators $\hat{\mathbf{J}} = \hat{\mathbf{L}} + \hat{\mathbf{S}}$. The Hamiltonian must now involve the interaction of the field and the total angular momentum. We shall not prove it, but in Russell–Saunders coupling,

$$\hat{H}_{\text{mag}} = g(e/2m_e)\hat{\mathbf{J}} \cdot \mathbf{B}$$

where

$$g = 1 + \frac{J(J+1) - L(L+1) + S(S+1)}{2J(J+1)}$$

is the Landé g-factor.

In this coupling scheme a state will be described by a ket $|j, m\rangle$ where, for convenience we omit the subscript on m. We have

$$\hat{H}_{\text{mag}}|j, m\rangle = g(e/2m_e)\hat{\mathbf{J}} \cdot \mathbf{B}|j, m\rangle$$
$$= g(e/2m_e)\hat{J}_z|j, m\rangle B_z$$

if the field is restricted to the z-direction. Thus using

Fig. 7.2. A typical splitting pattern in the Zeeman effect.

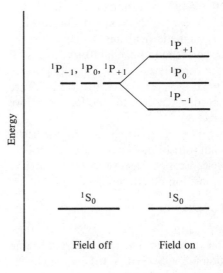

Fig. 7.3. The Zeeman effect for P terms (*a*) With the magnetic field on the number of spectral lines increases. (*b*) The $\Delta m = 0$ line has a different polarisation from the $\Delta m = \pm 1$ lines. (See text.)

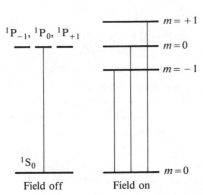

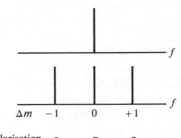

(*a*)   (*b*)

our rules in section 7.2,

$$\hat{H}_{\text{mag}}|j,m\rangle = \frac{gemh}{2m_e}|j,m\rangle$$

$$= g\beta m|j,m\rangle$$

where β is the Bohr magneton.

Because $\hat{H}_{\text{mag}}$ is the energy operator,

$$\hat{H}_{\text{mag}}|j,m\rangle = E_m|j,m\rangle$$

so

$$E_m = g\beta m.$$

This gives us the energies of the states $|j,m\rangle$ in the presence of the field. For example, a term $^1S_0$ has just one level $^1S_0$ for which $j = m = 0$. Our formula shows that a magnetic field has no effect on the energy of the corresponding state $|0,0\rangle$. For a ^1P term there are three levels: $^1P_{-1}$, $^1P_0$, $^1P_1$ which correspond to $|1,-1\rangle, |1,0\rangle, |1,1\rangle$. In fig. 7.2 are shown the splitting patterns for the four terms we have discussed.

Now we have the splitting pattern it is useful to find out what transitions can occur when the field is on. For an electric dipole transition there has to be a non-zero transition dipole moment, i.e.

$$\langle j_2, m_2|\hat{\mathbf{d}}|j_1, m_1\rangle \neq 0$$

where $|j_1, m_1\rangle$ and $|j_2, m_2\rangle$ are the two states involved in the transition. The selection rule, to join those in section 5.2, is $\Delta m = 0, \pm 1$. This gives us the lines shown in fig. 7.3(a). A feature of Zeeman spectra is that the light which is emitted is polarised. A transition with $\Delta m = -1$ gives rise to left-handed circularly polarised light, $\Delta m = +1$ to right-handed circularly polarised light. When $\Delta m = 0$ the light is plane polarised with its electric vector in the z-direction. This is illustrated in fig. 7.3(b) with $\Delta m = \pm 1$ transitions marked as σ, and $\Delta m = 0$ as π.

Questions

7.4   The sodium D-lines occur for transitions between ^2S and ^2P terms. See question 5.8.

(i) What is the cause of the two D-lines? Draw the energy level diagram.

(ii) What are the $|j,m\rangle$ states that can occur for the ^2S and ^2P terms?

(iii) Show the splitting of these states in the presence of a magnetic field.

(iv) Show the allowed transitions on your diagram.

(v) How do the spectra compare in the absence and presence of the magnetic field?

(vi) What will be the polarisation of the lines? This is an example of the anomalous Zeeman effect. The Zeeman effect occurs for singlet states and the 'anomalous' version for non-singlet states. Of course the effect is no longer regarded as anomalous.

7.4   Nuclear magnetic resonance

In n.m.r. spectroscopy the nuclear spin states have their degeneracies lifted by an applied magnetic field. Suppose we have a single free proton. Its energy of interaction with the field, **B**, will be given by the Hamiltonian

$$\hat{H} = -g\left(\frac{e}{2m_p}\right)\hat{\mathbf{I}}\cdot\mathbf{B}$$

where g is the proton g-value. (Note the opposite sign to the analogous expression for an electron.) $\hat{\mathbf{I}}$ is the nuclear spin operator.

If now we place the proton in a molecular environment we know that the field that it experiences is not **B**. Owing to the local field it is $(1-\sigma)\mathbf{B}$. Then,

$$\hat{H} = -g(e/2m_p)(1-\sigma)\hat{\mathbf{I}}\cdot\hat{\mathbf{B}}.$$

We shall write this as

$$\hat{H} = -k\hat{\mathbf{I}}\cdot\mathbf{B}.$$

If there are two protons, to a first approximation

$$\hat{H} = -k_1\hat{\mathbf{I}}_1\cdot\mathbf{B} - k_2\hat{\mathbf{I}}_2\cdot\mathbf{B}$$

or

$$\hat{H} = -k_1\hat{I}_{1z}B_z - k_2\hat{I}_{2z}B_z$$

for a field in the z-direction. The spin states for two protons can be written $|\alpha,\alpha\rangle, |\alpha,\beta\rangle, |\beta,\alpha\rangle, |\beta,\beta\rangle$. It is easy to show that these are eigenstates of the Hamiltonian with energies $-(k_1+k_2)$, $-(k_1-k_2)$, (k_1-k_2), (k_1+k_2) each in units of $\frac{1}{2}\hbar B_z$. If we assume $k_1 > k_2$ we obtain the splitting pattern of fig. 7.4. We expect to see just two lines in the spectrum, each being produced when one of the protons changes its spin state. Real spectra do of course show more than two lines. This is because of the interaction between the spins.

This is the spin–spin coupling, which can be written $(1/\hbar)J\hat{\mathbf{I}}_1\cdot\hat{\mathbf{I}}_2$ and occurs quite irrespective of the field. Here J is the spin–spin coupling constant; it has nothing to do with angular momentum. Now the Hamiltonian becomes

$$\hat{H} = -k_1\hat{I}_{1z}B_z - k_2\hat{I}_{2z}B_z + \frac{1}{\hbar}J\hat{\mathbf{I}}_1\cdot\hat{\mathbf{I}}_2.$$

But

$$\hat{\mathbf{I}}_1\cdot\hat{\mathbf{I}}_2 = \hat{I}_{1x}\hat{I}_{2x} + \hat{I}_{1y}\hat{I}_{2y} + \hat{I}_{1z}\hat{I}_{2z}$$

which can be rearranged to give

$$\hat{\mathbf{I}}_1\cdot\hat{\mathbf{I}}_2 = \tfrac{1}{2}(\hat{I}_{1+}\hat{I}_{2-} + \hat{I}_{1-}\hat{I}_{2+} + 2\hat{I}_{1z}\hat{I}_{2z}).$$

If we look at the effect of this operator on $|\alpha,\alpha\rangle$ we find, taking the terms one by one,

$$\hat{\mathbf{I}}_1\cdot\hat{\mathbf{I}}_2|\alpha,\alpha\rangle = \tfrac{1}{2}[0 + 0 + 2(\tfrac{1}{2}\hbar)(\tfrac{1}{2}\hbar)]$$
$$= \tfrac{1}{4}\hbar^2|\alpha,\alpha\rangle.$$

This follows because the operators $\hat{I}_{1+}$ and $\hat{I}_{2+}$ on $|\alpha,\alpha\rangle$ wipe out the kets: we cannot go over the top of the spin ladder.

$|\alpha,\alpha\rangle$ is therefore an eigenstate of $\hat{\mathbf{I}}_1\cdot\hat{\mathbf{I}}_2$. However, we can show by the same method that $|\alpha,\beta\rangle$ and $|\beta,\alpha\rangle$ are not such eigenstates. For example,

$$\hat{\mathbf{I}}_1\cdot\hat{\mathbf{I}}_2|\alpha,\beta\rangle = \tfrac{1}{2}\hbar^2|\beta,\alpha\rangle - \tfrac{1}{4}\hbar^2|\alpha,\beta\rangle$$

Fig. 7.4. In the absence of spin–spin coupling, an n.m.r. spectrum has just two lines. The energy of transitions marked 1 is $2k_2$, that of those marked 2 is $2k_1$. The energy is in units of $\tfrac{1}{2}\hbar B_z$.

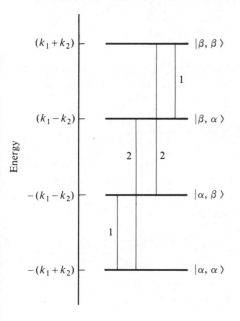

which shows that spin–spin coupling tends to mix $|\alpha,\beta\rangle$ and $|\beta,\alpha\rangle$. A result which follows from this is that combinations like $(1/\sqrt{2})(|\alpha,\beta\rangle + |\alpha,\beta\rangle)$ and $(1/\sqrt{2})(|\alpha,\beta\rangle - |\beta,\alpha\rangle)$ can occur. There is a corresponding change in the energy level diagram because the energies of all the states depend on the value of the spin–spin coupling constant. Thus the spectra become correspondingly complex. Also, not all lines have the same intensity. These points are revealed in the answers to the following questions.

Questions

7.5 The intensity of a transition is proportional to the square of the transition dipole moment.
  (i) Show why a transition between $|\alpha,\alpha\rangle$ and $\cos\theta|\alpha,\beta\rangle + \sin\theta|\beta,\alpha\rangle$ is proportional to $1 + \sin 2\theta$.
  (ii) If $\cos\theta|\alpha,\beta\rangle + \sin\theta|\beta,\alpha\rangle$ is one of the mixed states, why is the other $\sin\theta|\alpha,\beta\rangle - \cos\theta|\beta,\alpha\rangle$?
  (iii) Thus, which other transition has intensity proportional to $1 + \sin 2\theta$?
  (iv) The other two transitions have intensities proportional to $1 - \sin 2\theta$. Sketch the form of the pattern of intensities if $\theta = \pi/8$ (22.5°).

7.5 Perturbation theory

We have spent some time discussing the Zeeman effect and n.m.r. spectroscopy. Both phenomena are due to the effects of an applied magnetic field has on the energies of electrons and protons. The magnetic field perturbs their energies. The effect that a perturbation like this has can be accounted for in a quite general way. This is the role of perturbation theory.

If $\hat{H}_0$ is the Hamiltonian in the absence of the perturbation and $|n\rangle$ is its set of eigenstates, we know that

$$\hat{H}_0|n\rangle = E_n{}^0|n\rangle.$$

Here we have added a '0' as a subscript or superscript to show that there is no perturbation. With a perturbation present the Hamiltonian will have another term, $\hat{H}_1$ say. The total Hamiltonian is then

$$\hat{H} = \hat{H}_0 + \hat{H}_1.$$

Similarly, we could add $\hat{H}_2$, $\hat{H}_3$,... if there are further perturbations.

This is where the main assumption of perturbation theory enters. It is that the eigenstates of $\hat{H}$ can still be written as a combination of the states $|n\rangle$ which were the eigenstates in the absence of the perturbation. With this assumption we find in M.7.2 that the total energy is, to first order,

$$E = E_n^0 + \langle n|\hat{H}_1|n\rangle$$

or

$$E = E_0 + E_1 \text{ for short.}$$

To see how the theory works in a particular case, let us consider the Zeeman effect in the plane rigid rotator. We know that, in the absence of the field the Hamiltonian, $\hat{H}_0$ is

$$\hat{H}_0 = \hat{L}_z^2/2I.$$

The Hamiltonian for the interaction of a magnetic moment with a magnetic field is

$$\hat{H}_1 = -\hat{\mathbf{m}} \cdot \mathbf{B}$$

or

$$\hat{H}_1 = \frac{e}{2m_e} \hat{L}_z B_z$$

if we keep the field to the z-direction.
We have

$$\hat{H}_0|m\rangle = E_m|m\rangle$$

and we know from section 7.2 that $E_m = m^2\hbar^2/2I$. Thus, from perturbation theory,

$$E = \frac{m^2\hbar^2}{2I} + \langle m|\hat{H}_1|m\rangle$$

$$= \frac{m^2\hbar^2}{2I} + \frac{e}{2m_e}\langle m|\hat{L}_z|m\rangle B_z$$

or

$$E = \frac{m^2\hbar^2}{2I} + \frac{e\hbar m}{2m_e} B_z$$

because

$$\hat{L}_z|m\rangle = m\hbar|m\rangle.$$

The theory predicts that the perturbation will lift the degeneracy of the levels because new states with positive values of m will have different energies to those with negative values of m.

Questions

7.6 Sketch the energy level diagram for a plane rigid rotator in the absence and presence of a

magnetic field. Show the levels up to $|m| = 3$. Show the transitions allowed between the different levels and sketch the appearance of the spectrum. The selection rule is $\Delta m = \pm 1$.

7.7 For an atom which has its electrons obeying Russell–Saunders coupling, the states can be written $|j, m\rangle$. In the presence of a magnetic field the Hamiltonian is

$$\hat{H}_1 = \frac{e}{2m_e} g\hat{J}_z B_z$$

where g is the Landé g-factor

$$g = 1 + \frac{J(J+1) - L(L+1) + S(S+1)}{2J(J+1)}.$$

(i) What are the possible states when $j = \frac{5}{2}$? This could correspond to a term $^2D_{5/2}$ for example.
(ii) What does perturbation theory predict for the energies of the states $|\frac{5}{2}, m\rangle$? Sketch the splitting pattern.

7.8 Suppose a particle in a one-dimensional box is subject to a perturbation

$$\hat{H}_1 = \frac{\pi h^2}{64ml^2} \sin\left(\frac{\pi x}{l}\right).$$

Changing to the Schrödinger representation, the correction to the energy is E_1 where

$$E_1 = \int_0^l \psi^*(x)\hat{H}_1\psi(x)\,dx.$$

(i) Using $\psi = \sqrt{(2/l)} \sin(n\pi x/l)$, evaluate E_1.
(ii) Show that the new set of energy levels is given by

$$E_n = \left(1 + \frac{1}{4n^2 - 1}\right)\frac{n^2h^2}{8ml^2}.$$

(iii) What is the percentage change in the energy when $n = 1$ and when $n = 50$? Hint: to help with the integration use the identity

$$\sin^2 kx \sin ax = \tfrac{1}{4}[\sin(2k - a)x$$
$$- \sin(2k + a)x + 2\sin ax]$$

7.9 Perturbation theory is a very useful method of giving approximate solutions of Schrödinger's equation. There is another method which can also give approximate solutions. This is the variation method. To use the method we shall first prove the variation theorem. Let the true eigenvectors of the Hamiltonian be $|N_i\rangle$ such that $\hat{H}|N_i\rangle = E_i|N_i\rangle$. Now take a trial state $|T\rangle$ whose form we shall guess, and assume that $|T\rangle = \sum c_i|N_i\rangle$.

Let E_0 be the lowest eigenvalue of $\hat{H}$. Then,

$$\langle T|\hat{H} - E_0|T\rangle = \langle T|\hat{H}|T\rangle - E_0\langle T|T\rangle.$$

We leave it as part of the question to prove that this means

$$\langle T|\hat{H}|T\rangle - E_0 = \sum c_i^* c_i (E_i - E_0).$$

By hypothesis $E_i \geqq E_0$, and $c_i^* c_i$ can never be negative. Therefore,

$$\langle T|\hat{H}|T\rangle - E_0 \geqq 0$$

or

$$\langle T|\hat{H}|T\rangle \geqq E_0.$$

This proves the variation theorem: the energy of a trial state is always greater than or equal to the true energy.

We shall now use the theorem to estimate the lowest energy of a simple harmonic oscillator.

(i) Let $\psi_T = Ae^{-Cx^2}$ be our trial wavefunction. Determine the value of A by normalising ψ_T i.e. ensure that

$$\int_{-\infty}^{\infty} \psi_T^* \psi_T \, dx = 1.$$

(ii) Write down the Hamiltonian for the oscillator and show all the terms in $E_T = \int_{-\infty}^{\infty} \psi_T^* \hat{H} \psi_T \, dx$.

(iii) Evaluate each term in the integral.

(iv) To find the minimum energy, set $\partial E_T/\partial C = 0$ and calculate the necessary value for C.

(v) The frequency, f_0, is given by $f_0 = (1/2\pi)\sqrt{(k/m)}$. What is your value for the minimum energy? How does it compare with the true result?

Hints: See appendix A for the integrals. Note that if F is an even function, $\int_{-\infty}^{\infty} F \, dx = 2\int_0^{\infty} F \, dx$.

7.10 The variation theorem can be applied to the problem of obtaining the wavefunctions and energies of molecular orbitals. Suppose that we form a trial wavefunction, ψ_T, for a homopolar diatomic molecule

$$\psi_T = c_A\phi_A + c_B\phi_B.$$

Let the Hamiltonian for the molecule be $\hat{H}$. Then we can write the energy as

$$E_T = \frac{\int \psi_T^* \hat{H} \psi_T \, dv}{\int \psi_T^* \psi_T \, dv}$$

or as $E_T \int \psi_T^* \psi_T \, dv = \int \psi_T^* \hat{H} \psi_T \, dv$.

(i) Show, and explain why, this equation reduces to

$$E_T(c_A^2 + c_B^2 + 2c_A c_B S)$$
$$= c_A^2 \alpha + 2c_A c_B \beta + c_B^2 \alpha.$$

(See chapter 4 if you have forgotten the notation.)

(ii) If we now differentiate each side of this equation with respect to c_A we find that

$$\frac{\partial E_T}{\partial c_A}(c_A^2 + c_B^2 + 2c_A c_B S) E_T(2c_A + 2c_B S)$$
$$= 2c_A \alpha + 2c_B \beta.$$

Write down the analogous equation obtained by differentiating with respect to c_B.

(iii) In order to minimise the energy we must have $(\partial E_T/\partial c_A) = 0$ and $(\partial E_T/\partial c_B) = 0$. Show that this means that

$$c_A(\alpha - E_T) + c_B(\beta - E_T S) = 0$$
and
$$c_A(\beta - E_T S) + c_B(\alpha - E_T) = 0.$$

(iv) Now return to M.3.5 to show that for these simultaneous equations to have solutions,

$$\begin{vmatrix} (\alpha - E_T) & (\beta - E_T S) \\ (\beta - E_T S) & (\alpha - E_T) \end{vmatrix} = 0.$$

Expand the determinant to prove that either

$$E_T = \frac{\alpha + \beta}{1 + S} \quad \text{or} \quad E_T = \frac{\alpha - \beta}{1 - S}.$$

(v) Use these results and the simultaneous equations of (iii) to show that $c_1 = \pm c_2$.

(vi) Finally, by normalising ψ_T obtain the expressions for the bonding and antibonding orbitals.

7.6   The simple harmonic oscillator problem

By using the techniques we developed in the work on angular momentum operators we can derive the correct expression for the zero point energy of a simple harmonic oscillator. The method illustrates the power of the operator approach.

The Hamiltonian is

$$\hat{H} = \frac{\hat{p}_x^{\,2}}{2m} + \tfrac{1}{2}kx^2.$$

For reasons which will become apparent shortly, it is useful to put $\hat{H}$ in terms of two operators $\hat{A}_+$ and $\hat{A}_-$. Although not identical to $\hat{L}_+$ and $\hat{L}_-$, they have similar mathematical properties. We shall put

$$\hat{A}_+ = \frac{\hat{p}_x}{a} + iax; \quad \hat{A}_- = \frac{\hat{p}_x}{a} - iax$$

where $a = (km)^{1/4}$. The reason for these apparently perverse definitions is that

$$\hat{H} = \tfrac{1}{2}\sqrt{(k/m)}(\hat{A}_+\hat{A}_- + \hbar).$$

This is proved in M.7.3.
Alternatively,

$$\hat{H} = \tfrac{1}{2}\sqrt{(k/m)}(\hat{A}_-\hat{A}_+ - \hbar).$$

If

$$\hat{H}|n\rangle = E_n|n\rangle$$

we have from the first expression for $\hat{H}$,

$$E_n = \tfrac{1}{2}\sqrt{(k/m)}\langle n|\hat{A}_+\hat{A}_-|n\rangle + \hbar\langle n|n\rangle$$

or

$$\langle n|\hat{A}_+\hat{A}_-|n\rangle = 2\sqrt{(m/k)}E_n + \hbar$$

owing to the normalisation of $|n\rangle$.
Similarly,

$$\langle n|\hat{A}_-\hat{A}_+|n\rangle = 2\sqrt{(m/k)}E_n + \hbar.$$

We have seen before that combinations like $\langle n|\hat{A}_+\hat{A}_-|n\rangle$ are of the form $\langle S|S\rangle$ where $|S\rangle = \hat{A}_-|n\rangle$. Thus $\langle n|\hat{A}_+\hat{A}_-|n\rangle$ can never be negative, and we have

$$\langle n|\hat{A}_+\hat{A}_-|n\rangle \geq 0$$

and

$$\langle n|\hat{A}_-\hat{A}_+|n\rangle \geq 0.$$

Using our previous results,

$$2\sqrt{(m/k)}E_n - \hbar \geq 0 \tag{C}$$

and

$$2\sqrt{(m/k)}E_n + \hbar \geq 0.$$

By a similar argument to that in section 7.2, this means that there has to be a minimum energy. Let us call it E_0. From equation (C) we have

$$E_0 = \frac{h}{4\pi}\sqrt{\left(\frac{k}{m}\right)}.$$

But the theory of the simple harmonic oscillator shows that

$$k = 4\pi^2 mf^2.$$

Therefore

$$E_0 = \tfrac{1}{2}hf$$

which is the expression for the zero point energy that we sought to obtain.

By further applications of the operators $\hat{A}_+$ and $\hat{A}_-$ it is possible to prove that the energy levels are equally spaced by hf. However this is a somewhat laborious procedure which we will not pursue.

7.7   Summary

This, our last chapter, has displayed some of the power of quantum theory. Of necessity the examples we have seen have been somewhat arbitrarily chosen, but they do show something of the quantum chemist's armoury.

It is a measure of the success of quantum theory that it has insinuated itself into so much of chemistry that we may often take it for granted. For example, chemists will talk of orbitals as if they are 'really there' rather than mathematical abstractions that serve as useful models of the behaviour of electrons. It is wise to remember that quantum chemistry *is* concerned with making models. Such models can be good, bad, or indifferent. We can but try to match them with the results of experiments, and to make them as elegant as possible.

M.7.1

We wish to show that

$$[\hat{L}_z, \hat{L}_+] = \hbar \hat{L}_+.$$

Now

$$
\begin{aligned}
[\hat{L}_z, \hat{L}_+] &= \hat{L}_z \hat{L}_+ - \hat{L}_+ \hat{L}_z \\
&= \hat{L}_z(\hat{L}_x + i\hat{L}_y) - (\hat{L}_x + i\hat{L}_y)\hat{L}_z \\
&= \hat{L}_z \hat{L}_x - \hat{L}_x \hat{L}_z + i(\hat{L}_z \hat{L}_y - \hat{L}_y \hat{L}_z) \\
&= [\hat{L}_z, \hat{L}_x] + i[\hat{L}_z, \hat{L}_y].
\end{aligned}
$$

In question 6.10 we saw that

$$[\hat{L}_z, \hat{L}_x] = i\hbar \hat{L}_y \quad \text{and} \quad [\hat{L}_y, \hat{L}_z] = i\hbar \hat{L}_x.$$

Thus,

$$
\begin{aligned}
[\hat{L}_z, \hat{L}_+] &= i\hbar \hat{L}_y + \hbar \hat{L}_x \\
&= \hbar(\hat{L}_x + i\hat{L}_y) \\
&= \hbar \hat{L}_+.
\end{aligned}
$$

Similarly we can show that $[\hat{L}_z, \hat{L}_-] = \hbar \hat{L}_-$.

M.7.2

We shall make a small change to the equation for the energy in the absence of a perturbation:

$$\hat{H}_0 |n_0\rangle = E_n^0 |n_0\rangle.$$

Now the states in the absence of the perturbation are also given the '0' subscript. Now suppose that

$$\hat{H} = \hat{H}_0 + \lambda \hat{\mathscr{H}}_1$$

where λ is a small number and $\lambda \hat{\mathscr{H}}_1 \equiv \hat{H}_1$. Similarly, we can write the new states $|n\rangle$ in the presence of the perturbation as

$$|n\rangle = |n_0\rangle + \lambda |n_1\rangle$$

λ is the small number again and $|n\rangle_1$ represents the first order correction to the ground state $|n\rangle_0$. Following a similar notation we shall write

$$E_n = E_n^0 + \lambda E_n^1$$

where

$$\hat{H}|n\rangle = E_n |n\rangle.$$

If we replace each of these terms by those above we have

$$(\hat{H}_0 + \lambda \hat{\mathscr{H}}_1)(|n_0\rangle + \lambda |n_1\rangle) = (E_n^0 + \lambda E_n^1)(|n_0\rangle + \lambda |n_1\rangle),$$

i.e.

$$
\begin{aligned}
\hat{H}_0 |n_0\rangle &+ \lambda(\hat{H}_0 |n_1\rangle + \hat{\mathscr{H}}_1 |n_0\rangle) + \lambda^2 \hat{\mathscr{H}}_1 |n_1\rangle \\
&= E_n^0 |n_0\rangle + \lambda(E_n^0 |n_1\rangle \\
&\quad + E_n^1 |n_0\rangle) + \lambda^2 E_n^1 |n_1\rangle.
\end{aligned}
$$

For this relation to hold the terms in individual powers of λ must be equal. For the terms independent of λ,

$$\hat{H}_0 |n_0\rangle = E_n^0 |n_0\rangle;$$

and for those to the first power of λ

$$\hat{H}_0 |n_1\rangle + \hat{\mathscr{H}}_1 |n_0\rangle = E_n^0 |n_1\rangle + E_n^1 |n_0\rangle.$$

We cannot equate $\hat{\mathscr{H}}_1 |n_1\rangle$ with $E_n^1 |n_1\rangle$ because there are many terms in λ^2 that we have omitted. For example we should include $E_n = E_n^0 + \lambda E_n^1 + \lambda^2 E_n^2$. The first of our two equations is clearly the eigenvalue equation in the absence of the perturbation.

To simplify the second we make use of the basic assumption that the states $|n\rangle_1$ can be written as a combination of the states in the absence of the perturbation. Thus we employ the principle of super-position to put

$$|n_1\rangle = \sum c_j |j_0\rangle$$

so that

$$
\begin{aligned}
\hat{H}_0 \sum c_j |j_0\rangle + \hat{\mathscr{H}}_1 |n_0\rangle &= E_n^0 \sum c_j |j_0\rangle \\
&\quad + E_n^1 |n_0\rangle.
\end{aligned}
$$

If we act on both sides of the equation with $\langle n_0|$ we find

$$
\begin{aligned}
\sum c_j \langle n_0|\hat{H}_0|J_0\rangle + \langle n_0|\hat{\mathscr{H}}_1|n_0\rangle \\
= \sum c_j E_n^0 \langle n_0|j_0\rangle + E_n^1 \langle n_0|n_0\rangle.
\end{aligned}
$$

But $H_0|j_0\rangle = E_j^0|j_0\rangle$ so that

$$
\begin{aligned}
\sum c_j E_j^0 \langle n_0|j_0\rangle + \langle n_0|\hat{\mathscr{H}}_1|n_0\rangle \\
= \sum c_j E_n^0 \langle n_0|j_0\rangle + E_n^1 \langle n_0|n_0\rangle.
\end{aligned}
$$

However, $\langle n_0|j_0\rangle = \delta_{n,j}$ and $\langle n_0|n_0\rangle = 1$, which gives $c_n E_n{}^0 + \langle n_0|\hat{H}_1|n_0\rangle = c_n E_n{}^0 + E_n{}^1$. This gives us our final result that

$$E_n{}^1 = \langle n_0|\mathscr{H}_1|n_0\rangle$$

so

$$\lambda E_n{}^1 = \langle n_0|\lambda\mathscr{H}_1|n_0\rangle$$

or, dropping the subscripts,

$$\lambda E_n{}^1 = \langle n|\hat{H}_1|n\rangle.$$

This says that the first order correction to the energy, $\lambda E_n{}^1$, is found by operating on the ground state eigenvectors with the Hamiltonian corresponding to the perturbation.

M.7.3

Remembering that we are dealing with operators

$$\hat{A}_+\hat{A}_- = \left(\frac{\hat{p}_x}{a} + iax\right)\left(\frac{\hat{p}_x}{a} - iax\right)$$

$$= \frac{\hat{p}_x{}^2}{a} + a^2x^2 - i(\hat{p}_x x - x\hat{p}_x).$$

But

$$[\hat{p}_x, \hat{x}] = \frac{\hbar}{i},$$

so

$$\hat{A}_+\hat{A}_- = \frac{p_x{}^2}{a} + a^2x^2 - \hbar$$

$$= \frac{1}{\sqrt{(km)}}\hat{p}_x{}^2 + \sqrt{(km)}x^2 - \hbar.$$

Thus

$$\frac{1}{2}\sqrt{(k/m)}\hat{A}_+\hat{A}_- + \hbar = \frac{p_x{}^2}{2m} + kx^2$$

$$= \hat{H},$$

as required.

Answers to questions

$M \times LT^{-1} \times L = ML^2T^{-1}$. Thus angular momentum and action have the same units.

1.5 Using $\mathcal{T} = hf - \phi$,

$$\mathcal{T} = 6.626 \times 10^{-34} \times 10^{15}$$
$$- 3.68 \times 10^{-19} = 2.946 \times 10^{-19}\,\text{J}$$

$$v = \sqrt{\left(\frac{2\mathcal{T}}{m}\right)} = \sqrt{\left(\frac{2 \times 2.946 \times 10^{-19}}{9.109 \times 10^{-31}}\right)}$$

$$\approx 8 \times 10^5\,\text{m}\,\text{s}^{-1}.$$

The answer would be the same.

1.6 A graph of kinetic energy against frequency has slope h and intercept $-\phi$ on the energy axis. Estimates from the graph are $h \approx 6.6 \times 10^{-34}\,\text{Js}$ and $\phi \approx 7.5 \times 10^{-19}\,\text{J}$. The accepted value of ϕ for copper is $7.69 \times 10^{-19}\,\text{J}$.

1.7 (i) Because $L = m_e vr$ we have $m_e v^2/r = L^2/m_e r^3$. Thus

$$\frac{L^2}{m_e r^3} = \frac{e^2}{4\pi\varepsilon_0 r^2}$$

or

$$r = \frac{4\pi\varepsilon_0 L^2}{e^2 m_e}.$$

Then

$$r_n = \frac{\varepsilon_0 h^2 n^2}{\pi e^2 m_e}.$$

(ii) $E = \frac{1}{2}\left(\frac{e^2}{4\pi\varepsilon_0 r}\right) - \frac{e^2}{4\pi\varepsilon_0 r} = -\frac{1}{2}\left(\frac{e^2}{4\pi\varepsilon_0 r}\right).$

(iii) $E_n = -\frac{1}{2}\left(\frac{e^2}{4\pi\varepsilon_0 r_n}\right) = -\frac{e^4 m_e}{8\varepsilon_0^2 h^2 n^2}.$

(iv) Yes – see the working in (ii) above.

1.8 $E_n = -\dfrac{e^4 m_e Z^2}{8\varepsilon_0^2 h^2 n^2}$, i.e. $E_n \propto -\dfrac{Z^2}{n^2}$.

1.9 The ground state energy is $E_1 = -2.179 \times 10^{-18}\,\text{J}$ or $E_1 = -1312.1\,\text{kJ}\,\text{mol}^{-1}$. When $n = \infty$, $E_n = 0$. The ionisation energy is the energy needed to remove the electron from E_1 to E_∞. This is $1313.1\,\text{kJ}\,\text{mol}^{-1}$. The agreement is excellent.

1.10 For helium, $Z = 2$ so the energy is four times greater ($E \propto Z^2$) i.e. about $5252\,\text{kJ}\,\text{mol}^{-1}$ is predicted.

Chapter 1

1.1 (i) $E_n = nhf$ gives $n = 2 \times 10^{-17}/(6.626 \times 10^{-34} \times 10^{15})$ so $n \approx 30$.

(ii) The energy is $\frac{1}{2}mv^2 = 18\,\text{J}$. With $f = 1\,\text{Hz}$ this gives $n \approx 3 \times 10^{34}$. This also implies that the difference in energy between $n = 3 \times 10^{34}$ and $n = 3 \times 10^{34} - 1$ is absolutely immeasurable.

1.2 The graph is discontinuous. Many paths, such as P_1 and P_2 cannot occur. In a classical world h would be zero. By letting $h \to 0$, quantum effects disappear.

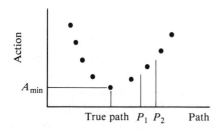

1.3 Planck's formula becomes $(8hf^3/c^3)\{1/[(1 + hf/kT) - 1]\}$ which is the classical formula $8f^2 kT/c^3$.

1.4 Action has dimension of (energy × time) i.e.

$$ML^2T^{-2} \times T = ML^2T^{-1}.$$

Angular momentum is given by mvr for a mass, m, moving in a circular path, radius r and with speed v. This gives us the units

1.11 Using $c = f\lambda$, we have $f = c/\lambda$ and $E = hc/\lambda$. Here,

$$E = \frac{6.626 \times 10^{-34} \times 2.998 \times 10^8}{10^{-9}}$$

$$= 1.986 \times 10^{-16}\,\text{J}.$$

It will take $2.179 \times 10^{-18}\,\text{J}$ to knock the electron out of the atom, thus leaving $1.965 \times 10^{-16}\,\text{J}$ to go into the kinetic energy. Therefore,

$$v = \sqrt{\left(\frac{2 \times 1.965 \times 10^{-16}}{9.109 \times 10^{-31}}\right)}$$

$$\approx 2.08 \times 10^8\,\text{m s}^{-1}.$$

Notice the similarity between this calculation and those on the photoelectric effect.

1.12 (i) $m_e v^2 = 2eV$, so $m_e^2 v^2 = 2m_e eV$, hence the result.

(ii) Energy $= 56.07\,\text{eV}$.
Using $n\lambda = 2d \sin\theta$ we have
$$\lambda = 2 \times 1.075 \times 10^{-10} \times \sin 75°$$
$$= 2.077 \times 10^{-10}\,\text{m}.$$

(iii) Using de Broglie's relation $h/\lambda = \sqrt{(2m_e eV)}$ or

$$\lambda = \frac{h}{\sqrt{(2m_e eV)}} = 2.073 \times 10^{-10}\,\text{m}.$$

The agreement is good.

1.13 For a golf ball put $m \approx 0.05\,\text{kg}$, for a car, $m \approx 1000\,\text{kg}$. Let them both have a speed of $40\,\text{m s}^{-1}$ (about 80 m.p.h.). Their de Broglie wavelengths would be of the order $3 \times 10^{-34}\,\text{m}$ and $3 \times 10^{-37}\,\text{m}$ respectively. We do not possess diffraction gratings of sufficient dimensions to show any effects associated with such small wavelengths.

1.14 $\lambda \approx 0.115\,\text{nm}$.

Chapter 2

2.1 (i) With $l = 0.56\,\text{nm}$, $E_1 = h^2/(8m_e l^2) = 1.921 \times 10^{-19}\,\text{J}$, $E_2 = 7.684 \times 10^{-19}\,\text{J}$, $E_3 = 17.289 \times 10^{-19}\,\text{J}$, $E_4 = 30.736 \times 10^{-19}\,\text{J}$, $E_5 = 48.025 \times 10^{-19}\,\text{J}$.

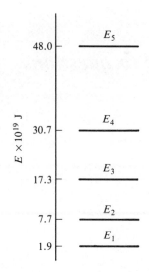

(ii) For E_1 to E_2, $\lambda \approx 576\,\text{nm}$; E_2 to E_3, $\lambda \approx 207\,\text{nm}$; E_3 to E_4, $\lambda \approx 148\,\text{nm}$; E_4 to E_5, $\lambda \approx 114\,\text{nm}$.

(iii) The E_2 to E_3 transition is the closest. The result shows that the free electron model can give the right order of magnitude for λ.

2.2 We have seven bonds to consider plus the two end contributions. This gives $l = 8 \times 0.14\,\text{nm} = 1.12\,\text{nm}$. For a transition between any two levels,

$$\Delta E = (n_2^2 - n_1^2)\frac{h^2}{8ml^2}$$

so

$$\lambda = \frac{8ml^2 c}{(n_2^2 - n_1^2)h}.$$

Putting in the values on both sides we find that $(n_2^2 - n_1^2) \approx 15$ i.e. $(n_2 - n_1)(n_2 + n_1) \approx 15$. Assuming transitions between adjacent levels only we find $n_2 = 8$, $n_1 = 7$ fits.

2.3 The model is simple: therein lies its attraction. For hydrocarbon chains it is surprisingly good. Physically it is unrealistic because the

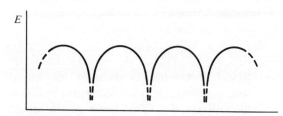

potential along the chain is not like that of a box. If looks more like this with wells coming at the nuclei of the atoms.

 2.4 $E_1 \approx 6 \times 10^{-20}$ J. Taking $m = 0.01$ kg and $l = 10$ m.

2.5 $E_1 \approx 5 \times 10^{-68}$ J, which is near enough to zero to make no difference in a large-scale, classical world.

2.6

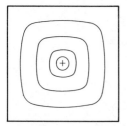

$n_x = n_y = 1$. $\psi(x, y)$
is everywhere positive.

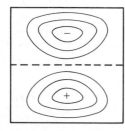

$n_x = 1$, $n_y = 2$
$\psi(x, y)$ has two lobes,
one positive, one negative.

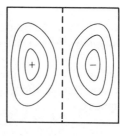

$n_x = 2$, $n_y = 1$

The dashed lines show the nodes.

2.7 (i)
$$-\frac{\hbar^2}{2m}\left(\frac{\partial^2 \psi}{\partial x^2} + \frac{\partial^2 \psi}{\partial y^2} + \frac{\partial^2 \psi}{\partial z^2}\right) = E\psi.$$

(ii)
$$E_{n_x, n_y, n_z} = (n_x^2 + n_y^2 + n_z^2)\frac{h^2}{8ml^2}.$$

(iii)
$$\psi(x, y, z)$$
$$= \left(\frac{2}{l}\right)^{3/2} \sin\frac{n_x \pi x}{l} \sin\frac{n_y \pi y}{l} \sin\frac{n_z \pi z}{l}.$$

2.8

E

Energy in units of $h^2/8ml^2$

4 ── (2, 0, 0), (0, 2, 0), (0, 0, 2)

3 ── (1, 1, 1)

2 ── (1, 1, 0), (1, 0, 1), (0, 1, 1)

1 ── (1, 0, 0), (0, 1, 0), (0, 0, 1)

For $E_{1,2,3}$ the energy is $14(h^2/8ml^2)$. We can obtain this with the following combinations of (n_x, n_y, n_z) (1, 2, 3), (1, 3, 2), (2, 1, 3), (2, 3, 1), (3, 1, 2), (3, 2, 1). The degeneracy is six-fold.

2.9 We have
$$\hat{H}\psi_- = \hat{H}[\psi_1(x, y) - \psi_2(x, y)]$$
$$= E_1\psi_1(x, y) - E_2\psi_2(x, y)$$
$$= E_1(\psi_1(x, y) - \psi_2(x, y)) \text{ if } E_1 = E_2$$
i.e.
$$\hat{H}\psi_- = E_1\psi_-$$
which proves the point.

For $\psi = a\psi_1 + b\psi_2$,
$$\hat{H}\psi = a\hat{H}\psi_1 + b\hat{H}\psi_2$$
$$= aE_1\psi_1 + bE_2\psi_2.$$

If $E_1 = E_2$,
$$\hat{H}\psi = E_1(a\psi_1 + b\psi_2) = E_1\psi.$$

2.10

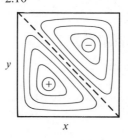

y

Contours of ψ_+

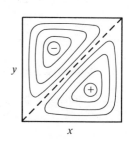

y

Contours of ψ_-

The contours are obtained by calculating values of

$$\psi_{\pm} = \frac{2}{l}\left[\sin\left(\frac{\pi x}{l}\right)\sin\left(\frac{2\pi y}{l}\right) \right. $$
$$\left. \pm \sin\left(\frac{2\pi x}{l}\right)\sin\left(\frac{\pi y}{l}\right) \right].$$

The diagonals shown dashed are the nodal lines.

2.11 Only $\psi_a(x)$ satisfies all the conditions. $\psi_1(x)$ disobeys (c); $\psi_2(x)$ disobeys (b); $\psi_3(x)$ disobeys (a).

2.12 The total probability must be one, therefore we must have the area under the graph of $|\psi(x)|^2$ equal to one i.e.

$$\int |\psi(x)|^2 dx = 1.$$

Taking the graphs in turn, $\psi_1(x)$ will not have a finite area as it does not come down towards the x-axis again. $\psi_2(x)$ has a discontinuity so its area is undefined. $\psi_3(x)$ is finite – it could certainly enclose unit area. $\psi_4(x)$ could have unit area, but in the region where it doubles over itself, $|\psi(x)|^2$ is many valued. In this region of x apparently we could obtain several different probabilities of finding a particle. This would not make physical sense.

Chapter 3

3.1 Use the formula for the 1s wavefunction from table 3.3. The program is trivial if you know any BASIC or FORTRAN. The contraction of the wavefunction with increasing nuclear change is *very* marked.

3.2 Let us form the combination

$$\tfrac{1}{2}(\Phi_1(\phi) + \Phi_{-1}(\phi)) = \frac{1}{2}\frac{1}{\sqrt{(2\pi)}}(e^{i\phi} + e^{-i\phi})$$

$$= \frac{1}{\sqrt{(2\pi)}}\cos\phi.$$

Similarly, the combination

$$\frac{1}{2i}(\Phi_1(\phi) - \Phi_1(\phi)) = \frac{1}{2i}\frac{1}{\sqrt{(2\pi)}}(e^{i\phi} - e^{-i\phi})$$

$$= \frac{1}{\sqrt{(2\pi)}}\sin\phi.$$

For $\Phi_0(\phi)$ we have $\Phi_0(\phi) = 1/\sqrt{(2\pi)}$.

We are quite free to use these combinations because we know that as they are proper linear combinations they are perfectly satisfactory solutions of the Schrödinger equations. Many people prefer the real combinations but mathematically it is usually more elegant to use the complex exponential forms.

3.3 For a 2p orbital,

$$R(r) = \frac{1}{2\sqrt{6}}\left(\frac{1}{a_0}\right)^{3/2}\frac{r}{a_0}e^{-r/2a_0}$$

Thus,

$$\frac{d}{dr}r^2|R(r)|^2 = \frac{1}{24}\left(\frac{1}{a_0}\right)^5\frac{d}{dr}(r^4 e^{-r/a_0})$$

$$= \frac{1}{24}\left(\frac{1}{a_0}\right)^5\left[4r^3 - \frac{r^4}{a_0}\right]e^{-r/a_0}.$$

This is zero when $r = 0$ or $r = 4a_0$.

3.4 (i) In atomic units, the equation becomes

$$-\tfrac{1}{2}\nabla^2\psi - \frac{1}{r}\psi = E\psi.$$

(ii) $E_n = -\dfrac{1}{2}\dfrac{1}{n^2}.$

(iii) Notice that the requirement that $\int|\psi|^2 dv = 1$ means that $|\psi|^2 dv$ must be dimensionless. dv has the units of volume so $|\psi|^2$ has the reciprocal units. Ordinarily this is guaranteed by the factor $(1/a_0)^{3/2}$ in the wave functions. In the world of atomic units where a_0 is the unit of length, the unit of volume will be a_0^3. Scaling these to atomic units, just as a_0 disappears, so must a_0^3. Thus the 1s orbital is just $(1/\sqrt{\pi})e^{-r}$.

3.5 For a 2s orbital

$$r_{av} = \frac{1}{32\pi a_0^3}\int_0^\infty r^3\left(2 - \frac{r}{a_0}\right)^2 e^{-r/a_0}\,dr$$

$$\times \int_0^\pi \sin\theta\,d\theta\int_0^{2\pi} d\phi$$

$$= \frac{1}{8a_0^3}\int_0^\infty\left(4r^3 - \frac{4r^4}{a_0} + \frac{r^5}{a_0^2}\right)e^{-r/a_0}\,dr$$

$$= \frac{1}{8a_0^3}[24a_0^4 - 96a_0^4 + 120a_0^4]$$

so

$$r_{av} = 6a_0.$$

For a 2p orbital, the $2p_z$ for example,

$$r_{av} = \frac{1}{32\pi a_0{}^5} \int_0^\infty r^5 e^{-r/a_0} dr$$

$$\times \int_0^\pi \cos^2\theta \sin\theta \, d\theta \int_0^{2\pi} d\phi$$

$$= \frac{1}{16 a_0{}^5} 120 a_0{}^6 \int_0^\pi \cos^2\theta \sin\theta \, d\theta$$

$$= \frac{15a_0}{2} [-\tfrac{1}{3}\cos^3\theta]_0^\pi$$

or

$$r_{av} = 5a_0.$$

3.6 (i) The integral $\iint (2s\alpha)^*(2s\beta)ds d\upsilon$

$$= \int |2s|^2 d\upsilon \int \alpha^* \beta \, ds$$

$$= 0$$

because α^* and β are orthogonal.

(ii) These are orthogonal because 2s and $2p_z$ orbitals are orthogonal.

3.7 Because all electrons have $s = \tfrac{1}{2}$, thus we gain little from making it explicit.

3.8 See table 3.5.

3.9 $Cr: 1s^2 2s^2 2p^6 3s^2 3p^6 4s^2 3d^4$ predicted.
The fact that the outer two orbitals are $4s^1 3d^5$ suggests that too much energy is needed to put two electrons into the same 4s orbital. The energy is minimised by keeping only one electron in each orbital.

3.10

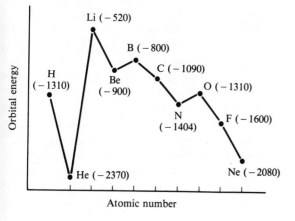

The graph is, of course, obtained by inverting fig. 3.12. The numbers are the (approximate) orbital energies in $kJ\,mol^{-1}$.

3.11

$$N_n{}^2 \int r^{2n} e^{-2\zeta r/a_0} dr = 1$$

so

$$N_n{}^2 \frac{(2n)!}{(2\zeta/a_0)^{2n+1}} = 1$$

or

$$N_n = \pm \left(\frac{2\zeta}{a_0}\right)^{(2n+1)/2} \left(\frac{1}{(2n)!}\right)^{1/2}.$$

3.12 The main difference is that the Slater 2s radial probability density does not become zero at $r = 2a_0$.

3.13 The combination $2s + k2p_z$ becomes

$$\frac{1}{4\sqrt{(2\pi)}}\left(\frac{1}{a_0}\right)^{3/2}\left[2 - \frac{r}{a_0} + \frac{kz}{a_0}\right]e^{-r/2a_0}$$

where $r = +\sqrt{(z^2)}$

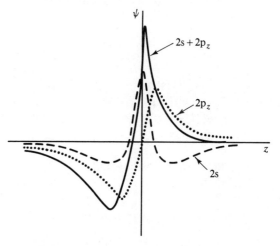

The symmetry of the 2s orbital is destroyed. It leans more heavily in the positive z-direction the more $2p_z$ character is included. As k increases, angular correlation increases.

3.14 (i) $\psi(1,2) = \dfrac{1}{\sqrt{6}} \begin{vmatrix} 1s\alpha(1) & 1s\alpha(2) & 1s\alpha(3) \\ 1s\beta(1) & 1s\beta(2) & 1s\beta(3) \\ 2s\alpha(1) & 2s\alpha(2) & 2s\alpha(3) \end{vmatrix}$

$$= \frac{1}{\sqrt{6}}\{1s\alpha(1)[1s\beta(2)2s\alpha(3)$$

$$- 2s\alpha(2)1s\beta(3)]$$

$$- 1s\alpha(2)[1s\beta(1)2s\alpha(3)$$
$$- 2s\alpha(1)1s\beta(3)]$$
$$+ 1s\alpha(3)[1s\beta(1)2s\alpha(2)$$
$$- 2s\alpha(1)1s\beta(2)]\}.$$

(ii) It is obtained by letting the third electron have β spin i.e. put $\phi_3 \equiv 2s\beta$.

(iii) The multiplicity is 2.

3.15 This can be achieved by the three stages
(i) add row 2 to row 1
(ii) subtract half of row 1 from row 2
(iii) extract the factor $-\frac{1}{2}$ from row 2.

3.16 (i) $\psi_+(r_1,r_2)\psi_-(r_1,r_2)$
$$= \frac{1}{2}\{[1s(r_1)2s(r_2)]^2$$
$$+ 1s(r_1)2s(r_2)1s(r_2)2s(r_1)$$
$$- 1s(r_1)2s(r_2)1s(r_2)2s(r_1)$$
$$- [1s(r_2)2s(r_1)]^2\}$$
$$= \frac{1}{2}\{[1s(r_1)2s(r_2)]^2$$
$$- [1s(r_2)2s(r_1)]^2\}.$$

Thus

$$\int\int \psi_+(r_1,r_2)\psi_-(r_1,r_2)dv_1 dv_2$$

$$= \frac{1}{2}\int|1s(r_1)|^2 dv_1 \int|2s(r_2)|^2 dv_2$$

$$- \frac{1}{2}\int|1s(r_2)|^2 dv_2 \int|2s(r_1)|^2 dv_1$$

$$= \frac{1}{2}(1 \times 1) - \frac{1}{2}(1 \times 1)$$

$$= 0.$$

Thus they are orthogonal.

(ii) When $r_1 = r_2$,

$$\psi_+(r_1,r_2) = \frac{2}{\sqrt{2}} 1s(r_1)2s(r_1)$$

$$\psi_-(r_1,r_2) = 0.$$

(iii) $\psi_-(r_1,r_2)$.

(iv) $\psi_+(r_1,r_2)$ occurs with the singlet spin functions and $\psi_-(r_1,r_2)$ with the triplet spin functions.

(v) The triplet state corresponds to the electrons having parallel spins. The spin functions for such states are associated with antisymmetric spatial wavefunctions like $\psi_-(r_1,r_2)$. Thus the notion of electrons with parallel spins avoiding

one another is really another manifestation of the Pauli principle.

(vi) Spin correlation is another way of describing the ramifications of the Pauli principle. It is not due to a 'force' in the sense that there is an electrostatic force of repulsion between two electrons.

3.17 (i) All the electrons could go into the lowest lying energy level. For example, the ground state electron configuration of oxygen would be $1s^8$.

(ii) These nuclear spin wavefunctions have exactly the same form as those for the electron spin functions of two electrons. The total wavefunction must not only be anti-symmetric to the exchange of any two electrons, but also to the exchange of any two protons (or neutrons). Thus the nuclear spin functions can either form a set of three symmetrical nuclear spin wave functions or one antisymmetric function. This is the basis for the appearance of ortho and parahydrogen in the ratio 3:1.

Chapter 4

4.1 F_2:$(1s\sigma_g)^2(1s\sigma_u{}^*)^2(2s\sigma_g)^2(2s\sigma_u{}^*)^2(2p_z\sigma_g)^2 \cdot$ $(2p_x\pi_u)^2(2p_y\pi_u)^2(2p_x\pi_g{}^*)^2(2p_y\pi_g{}^*)^2$. Bond order $= 1$. All the electrons are paired so fluorine is not paramagnetic.

4.2 N_2:$(1s\sigma_g)^2(1s\sigma_u{}^*)^2(2s\sigma_g)^2(2s\sigma_u{}^*)^2.$ $(2p_x\pi_u)^2(2p_y\pi_u)^2(2p_z\sigma_g)^2$. Note the reversal of the order of the $2p_x\pi_u$, $2p_y\pi_u$ and $2p_z\sigma_g$ orbitals. Bond order $= 3$. All the electrons are paired so nitrogen is not paramagnetic.

4.3 The outermost orbital for F_2^+ is $(2p_y\pi_g{}^*)^1$ and for N_2^+ it is $(2p_z\sigma_g)^1$. Both are paramagnetic owing to the unpaired electron. (i) The bond order for F_2^+ is now $1\frac{1}{2}$, and for N_2^+ is $2\frac{1}{2}$. Thus we would expect the bonding in F_2^+ to be stronger than in F_2. This agrees with the observed values of $150\,\text{kJ mol}^{-1}$ for the bond energy of F_2 and $270\,\text{kJ mol}^{-1}$ for F_2^+. (ii) Similarly the values $941\,\text{kJ mol}^{-1}$ for N_2 and $854\,\text{kJ mol}^{-1}$ for N_2^+. Support the view that the bonding in N_2^+ is weaker than in N_2.

4.4 The oxygen atom is more electronegative than carbon. Thus the electron distribution should be distorted towards the oxygen atom.

C   O

4.5   The method is exactly the same as M.4.4 except that a minus sign replaces the positive sign throughout.

4.6   For the combination $\phi_A + \phi_B$ we have

$$\int \phi_A \hat{H} \phi_A dv + \int \phi_A \hat{H} \phi_B dv + \int \phi_B \hat{H} \phi_A dv$$

$$+ \int \phi_B \hat{H} \phi_B dv = E(2 + 2S)$$

so

$$\alpha_A + \beta + \beta + \alpha_B = E(2 + 2S)$$

because the two bond integrals must be the same. Then

$$E = \frac{\alpha_A + \alpha_B + 2\beta}{2 + 2S}.$$

Similarly, for $\phi_A - \phi_B$,

$$E = \frac{\alpha_A + \alpha_B - 2\beta}{2 - 2S}.$$

4.7   (i) Normalisation of ψ means that

$$c_1{}^2 + c_2{}^2 + c_3{}^2 + c_4{}^2 = 1.$$

(ii) When $x = 1.62$

$$1.62c_1 + c_2 \qquad\qquad = 0 \quad (1)$$
$$c_1 + 1.62c_2 + c_3 \qquad = 0 \quad (2)$$
$$c_2 + 1.62c_3 + c_4 = 0 \quad (3)$$
$$c_3 + 1.62c_4 = 0 \quad (4)$$

From (1), $c_2 = -1.62c_1$.
From (4), $c_3 = -1.62c_4$.
Then (3) gives $c_2 = [(1.62)^2 - 1]c_4$.
Thus

$$c_1 = -\frac{[(1.62)^2 - 1]}{1.62} c_4.$$

Putting these into the normalisation condition gives

$$\left[\frac{(1.62)^2 - 1}{1.62} \right]^2 + [(1.62)^2 - 1]$$

$$+ (-1.62)^2 + 1 = \frac{1}{c_4{}^2}.$$

Then $c_4 \approx \pm 0.37$.
Taking the positive sign gives $c_3 \approx 0.60$, $c_2 \approx 0.60$, and $c_1 \approx 0.60$.

A similar calculation produces the values used in M.4.5 for ψ_2, ψ_3 and ψ_4.

4.8   Call the molecular orbital $\psi = c_1\phi_1 + c_2\phi_2$. Then operating with $\hat{H}$, multiplying by ϕ_1 and integrating,

$$c_1\alpha + c_2\beta = Ec_1$$

or

$$c_1(\alpha - E) + c_2\beta = 0.$$

Similarly we find $c_1\beta + c_2(\alpha - E) = 0$. These equations simplify to

$$xc_1 + c_2 = 0$$
$$c_1 + xc_2 = 0.$$

The secular equation is

$$\begin{vmatrix} x & 1 \\ 1 & x \end{vmatrix} = 0$$

or

$$x^2 - 1 = 0.$$

Thus

$$x = \pm 1.$$

Hence

$$E_1 = \alpha + \beta, \quad E_2 = \alpha - \beta$$

where $x = +1, c_1 + c_2 = 0$ which implies $c_1 = -c_2$; when $x = -1, -c_1 + c_2 = 0$ which implies $c_1 = +c_2$. Also, we must have $\int |\psi|^2 dv = 1$. This means that $c_1{}^2 + c_2{}^2 = 1$. Hence $c_1 = c_2 = \pm 1/\sqrt{2}$ and

$$\psi_1 = \frac{1}{\sqrt{2}}(\phi_1 + \phi_2); \quad \psi_2 = \frac{1}{\sqrt{2}}(\psi_1 - \psi_2).$$

The wavefunctions are just like those of fig. 4.12. ψ_2 is higher in energy than ψ_1.

4.9   (i) The secular equation is found to be

$$\begin{vmatrix} x & 1 & 0 & 0 & 0 & 1 \\ 1 & x & 1 & 0 & 0 & 0 \\ 0 & 1 & x & 1 & 0 & 0 \\ 0 & 0 & 1 & x & 1 & 0 \\ 0 & 0 & 0 & 1 & x & 1 \\ 1 & 0 & 0 & 0 & 1 & x \end{vmatrix} = 0.$$

(ii) On expanding the determinant we find

$$x^6 - 6x^4 + 9x^2 - 4 = 0.$$

Removing a factor $x^2 - 1$, once

$$(x^2 - 1)(x^4 - 5x^2 + 4) = 0.$$

Removing it again,

$$(x^2 - 1)(x^2 - 1)(x^2 - 4) = 0$$

or

$$(x - 1)(x + 1)(x - 1)(x + 1)$$
$$\times (x - 2)(x + 2) = 0.$$

Hence the roots are $x = \pm 1$ (twice), $x = \pm 2$. The energy level diagram is

(iii) and (iv)

$$E$$

—————— $E_6 = \alpha - 2\beta$

——— ——— $E_4, E_5 = \alpha - \beta$

⇅ ⇅ $E_2, E_3 = \alpha + \beta$

⇅ $E_1 = \alpha + 2\beta$

Note the degeneracy of E_2, E_3 and E_4, E_5.

(v) For convenience the wavefunctions are drawn out with the carbon atoms in a line; in reality the carbon will form a ring with contours 1 and 6 joined together.

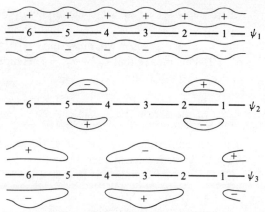

4.10 (i) For ethene $E = \alpha + \beta$.

(ii) For three such bonds, the total is $6\alpha + 6\beta$.

(iii) The six electrons will fill the levels E_1, E_2 and E_3 giving a total energy $2(\alpha + 2\beta) + 4(\alpha + \beta) = 6\alpha + 8\beta$.

(iv) According to Hückel the resonance energy is $6\alpha + 8\beta - (6\alpha + 6\beta) = 2\beta$.

4.11 $C_3H_3^+$, with $n = 0$, and $C_5H_5^-$ with $n = 1$. The σ framework in cyclic C_3 molecules is very strained and generally such molecules are only stable at low temperatures. $C_5H_5^-$ is well known to occur in 'sandwich' compounds like ferrocene, $Fe(C_2H_5)_2$.

4.12 Only the first three orbitals are filled by the six π electrons. Thus the coefficients for ψ_4, ψ_5 and ψ_6 are irrelevant.

Using the formula from M.4.7, for atom 1,

$$q_1 = 2\left(\frac{1}{\sqrt{6}}\right)^2 + 2\left(\frac{1}{2}\right)^2 + 2\left(\frac{1}{\sqrt{12}}\right)^2 \approx 1.$$

Similarly,

$$q_2 = q_3 = q_4 = q_5 = q_6 = 1.$$

We should expect this by virtue of the symmetry of benzene.

The bond order for the bond between atoms 1 and 2 is

$$P_{12} = n_1 c_{11} c_{12} + n_2 c_{21} c_{22} + n_3 c_{31} c_{32}$$
$$= 2\left(\frac{1}{\sqrt{6}}\right)\left(\frac{1}{\sqrt{6}}\right) + 2(\tfrac{1}{2})(\tfrac{1}{2})$$
$$+ 2\left(\frac{1}{\sqrt{12}}\right)\left(-\frac{1}{\sqrt{12}}\right)$$
$$= \tfrac{1}{3} + \tfrac{1}{2} - \tfrac{1}{6}$$

i.e.

$$P_{12} = \tfrac{2}{3}.$$

For atoms 2 and 3,

$$P_{23} = 2\left(\frac{1}{\sqrt{6}}\right)\left(\frac{1}{\sqrt{6}}\right) + 2\left(\frac{1}{2}\right)0$$
$$+ 2\left(-\frac{1}{\sqrt{12}}\right)\left(-\frac{2}{\sqrt{12}}\right)$$
$$= \tfrac{2}{3}.$$

Again, we find that the bond orders for all the bonds have the same value, $\tfrac{2}{3}$.

4.13

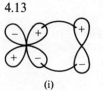

(i)

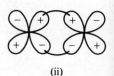

(ii)

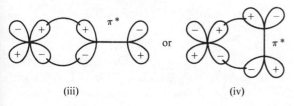

(iii)   (iv)

4.14 (i) With

$$4s: \frac{1}{\sqrt{4}}(\phi_1 + \phi_2 + \phi_3 + \phi_4)$$

(ii) $3d_{x^2-y^2}: \dfrac{1}{\sqrt{4}}(\phi_1 - \phi_2 + \phi_3 - \phi_4)$

(iii) $4p_x: \dfrac{1}{\sqrt{2}}(\phi_1 - \phi_2)$

(iv) $4p_y: \dfrac{1}{\sqrt{2}}(\phi_3 - \phi_4)$
} degenerate

4.15 The wavefunctions are

$$\psi_{III} = [\phi_2(2)\phi_5(5) + \phi_2(5)\phi_5(2)]$$
$$\cdot [\phi_3(3)\phi_4(4) + \phi_3(4)\phi_3(3)]$$
$$\cdot [\phi_1(1)\phi_6(6) + \phi_1(6)\phi_6(1)]$$

$$\psi_{IV} = [\phi_1(1)\phi_4(4) + \phi_1(4)\phi_4(1)]$$
$$\cdot [\phi_2(2)\phi_3(3) + \phi_2(3)\phi_3(2)]$$
$$\cdot [\phi_5(5)\phi_6(6) + \phi_5(6)\phi_6(5)]$$

$$\psi_V = [\phi_1(1)\phi_2(2) + \phi_1(2)\phi_2(1)]$$
$$\cdot [\phi_3(3)\phi_6(6) + \phi_3(6)\phi_6(3)]$$
$$\cdot [\phi_4(4)\phi_5(5) + \phi_4(5)\phi_5(4)].$$

(i) The simplest combination is

$$\psi_{Total} = \psi_I + \psi_{II} + \psi_{III} + \psi_{IV} + \psi_V.$$

(ii) The most general is

$$\psi_{Total} = c_1\psi_I + c_2\psi_{II} + c_3\psi_{III} + c_4\psi_{IV} + c_5\psi_V.$$

4.16 (i) The positions are $\mathbf{k}$ and $-\mathbf{k}$.
(ii) The combinations are p_z and $-p_z$.
(iii) Adding in the s orbital contribution we have

$$\psi_1 = s + p_z; \qquad \psi_2 = s - p_z.$$

(iv) Normalised, they are

$$\psi_1 = \frac{1}{\sqrt{2}}(s + p_z); \qquad \psi_2 = \frac{1}{\sqrt{2}}(s - p_z).$$

Incidentally, note that ψ_1 and ψ_2 are orthogonal.

(v) The shapes are

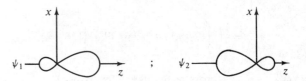

For the sp^2 hybrids, the position of the As are: $+\mathbf{k}$, $\mathbf{i}\cos 30 - \mathbf{k}\cos 60$, $-\mathbf{i}\cos 30 - \mathbf{k}\cos 60$ i.e.

$$\mathbf{k}, \qquad \frac{\sqrt{3}}{2}\mathbf{i} - \tfrac{1}{2}\mathbf{k}, \qquad -\frac{\sqrt{3}}{2}\mathbf{i} - \tfrac{1}{2}\mathbf{k}.$$

The orbitals are,

$$\psi_1 = s + p_z; \qquad \psi_2 = s + \frac{\sqrt{3}}{2}p_x - \tfrac{1}{2}p_z;$$

$$\psi_3 = s - \frac{\sqrt{3}}{2}p_x - \tfrac{1}{2}p_z.$$

When normalised they become

$$\psi_1 = \frac{1}{\sqrt{2}}(s + p_z);$$

$$\psi_2 = \frac{1}{\sqrt{2}}\left(s + \frac{\sqrt{3}}{2}p_x - \tfrac{1}{2}p_z\right);$$

$$\psi_3 = \frac{1}{\sqrt{2}}\left(s - \frac{\sqrt{3}}{2}p_z - \tfrac{1}{2}p_z\right).$$

(vi)

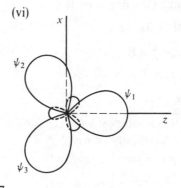

4.17

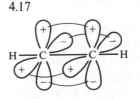

Ethyne

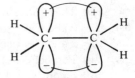

Ethene

4.18 $\int |\psi_1|^2 d\upsilon$

$$= \frac{1}{4}\left(\int |2s|^2 d\upsilon + \text{terms like } \int |2p_x|^2 d\upsilon \right.$$

$$\left. + \text{cross terms like } \int 2p_x 2p_y d\upsilon \right)$$

$$= \tfrac{1}{4}(1 + 3 + 0)$$

$$= 1$$

as required

4.19 If $\psi = N(s + kp)$ then $N^2 = 1 + k^2$ so $N = \pm(1 + k^2)$.

4.20 Molecular orbital theory is able to give an account of the bonding in methane. The tetrahedral sp^3 hybrids are derived using the fact that methane is tetrahedral. Therefore the tetrahedral arrangement of hybrids cannot be used to justify the validity of the theory.

4.21 (i) $\mathbf{i}, \mathbf{j}$, and $\mathbf{k}$ are odd.

(ii) The sp^3 hybrids the neither odd nor even. For example

$$I\psi_1 = I(2s + 2p_x + 2p_y + 2p_z)$$

$$= (2s - 2p_x - 2p_y - 2p_z)$$

which is neither $+\psi_1$ or $-\psi_1$.

Chapter 5

5.1 If we change the origin by $\mathbf{R}$

$$\boldsymbol{\mu} = \sum q_i(\mathbf{r}_i + \mathbf{R})$$

$$= \sum q_i \mathbf{r}_i + \sum q_i \mathbf{R}$$

$$= \sum q_i \mathbf{r}_i + \mathbf{R} \sum q_i$$

$$= \sum q_i \mathbf{r}_i$$

because $\mathbf{R}$ is a constant and because $\sum q_i = 0$.

5.2 $\mathbf{m}$ and $\mathbf{l}$ are antiparallel e.g. $\overset{\mathbf{l}}{\leftarrow} \overset{\mathbf{m}}{\rightarrow}$.

5.3 (i) $L = m_i vr = m_i r^2 \omega = m_i r^2 2\pi f$.

(ii) $\mu = \dfrac{Aq}{t} = \pi r^2 \dfrac{-e}{1/f} = -\pi r^2 ef$.

(iii) The ratio μ/L is therefore $-e/2m_e$.

5.4 $L = 0$ and $S = \tfrac{1}{2}$ so term is ^2S for Li.
$L = 2$ and $S = \tfrac{1}{2}$ so term is ^2D for Sc.

5.5 (i) Parallel. $S = 1$, multiplicity $= 3$, a triplet.

(ii) l_z can be 2, 1, 0, -1, -2. Cannot have both electrons in the same orbital so the

combinations are 3, 2, 1, 0, -1, -2, -3.

(iii) $L = 3$.

(iv) ^3F. Actually the level is $^3F_2$.

5.6 (i) To make L a maximum, we must have 2 electrons with $l_z = 2$, 2 with $l_z = 1$, 2 with $l_z = 0$ and 1 each with $l_z = -1$, $l_z = -2$.

Thus $L_z = 2 \times 2 + 2 \times 1 + 2 \times 0$
$$+ 1 \times (-1) + 1 \times (-2)$$
$$= 3.$$

As $S = 1$, the term is ^3F. Notice that this is the same as titanium $[Ar]4s^23d^2$. This illustrates the rule that one electron configuration nd^x gives rise to the same terms as nd^{10-x}.

(ii) $^3P < {}^3P_1 < {}^3P_2$.

(iii) $^3D_1 < {}^3D_2 < {}^3D_3$.

(iv) No. Hund's rules apply to the same electron configurations. The ^3P and ^3D levels arise from different ones.

(v)

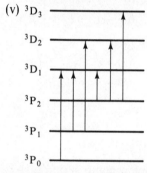

(vi) Transitions are shown by arrows on the diagram.

5.7 (i) $J_z = \tfrac{3}{2}, \tfrac{1}{2}, -\tfrac{1}{2}, -\tfrac{3}{2}$.

(ii) $J_z = \tfrac{1}{2}, -\tfrac{1}{2}$.

(iii) $L_z = 1, 0, -1$.

(iv) $S_z = +\tfrac{1}{2}, -\tfrac{1}{2}$.

(v) $\tfrac{3}{2}, \tfrac{1}{2}, \tfrac{1}{2}, -\tfrac{1}{2}, -\tfrac{1}{2}, -\tfrac{3}{2}$.

(vi) $J = \tfrac{3}{2}$ and $J = \tfrac{1}{2}$.

(vii) $^2P_{1/2}$, $^2P_{3/2}$.

5.8 (i) The levels are $^2S_{1/2}$, $^2P_{3/2}$, $^2P_{1/2}$.

(ii) The values of $E_{s.o}$ are, respectively, 0, $\tfrac{1}{2}\zeta_{3,1}$, $-\zeta_{3,1}$.

(iii)

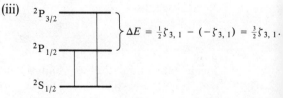

$$\Delta E = \tfrac{1}{2}\zeta_{3,1} - (-\zeta_{3,1}) = \tfrac{3}{2}\zeta_{3,1}.$$

(iv) In cm^{-1}, the frequencies of the lines are $(588.996 \times 10^{-7})^{-1}$ and $(589.593 \times 10^{-7})^{-1}$. The difference is $17.19\,cm^{-1}$. Thus $\xi_{3,1} = 11.46\,cm^{-1}$.

5.9 (i) $O_2:(1s\sigma_g)^2(1s\sigma_u{}^*)^2(2s\sigma_g)^2\,(2s\sigma_u{}^*)^2(2p_z\sigma_g)^2$ $(2p_x\pi_u)^2\,(2p_y\pi_u)^2(2p_x\pi_g{}^*)^1(2p_y\pi_g{}^*)^1$

(ii) The two electrons are in separate π orbitals with spins parallel. Remembering that for p orbitals, $m_l = +1$ for a p_x orbital, and $m_l = -1$ for a p_y orbital. By analogy, λ values for $2p_x\pi^*$ and $2p_y\pi^*$ must be ± 1, hence $\Lambda = 0$ and a Σ state results. The parallel spins result in a triplet state. As both π^* orbitals are g, they must give rise to a g state. That the state must have the term $^3\Sigma_g^-$ rather than $^3\Sigma_g^+$ can be seen by examining a sketch of π^* orbitals and their reflection in a mirror plane taken through the nuclei of the atoms.

5.10 According to the selection rules none of the transitions between these states is allowed. However, some transitions do actually occur. These transitions are all very much weaker than an allowed $^3\Sigma_u^- \leftarrow {}^3\Sigma_g^-$ transition which occurs at about 200 nm.

5.11

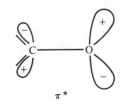

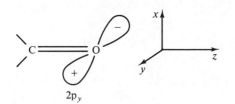

The transition $\pi^* \leftarrow 2p_y$ involves a rotation of charge so can be magnetic dipole allowed.

5.12 An electron entering a π^* orbital does what its name implies i.e. brings about a lowering of the strength of the bonding. The bond length therefore increases.

5.13 The transition dipole moment is

$$\int (s + kp)\boldsymbol{\mu}(s + kp)dv$$

$$= \int s\boldsymbol{\mu}s\,dv + k\int p\boldsymbol{\mu}s\,dv$$

$$+ k\int s\boldsymbol{\mu}p\,dv + k^2\int p\boldsymbol{\mu}p\,dv.$$

Of these, only $\int p\boldsymbol{\mu}s\,dv$ and $\int s\boldsymbol{\mu}p\,dv$ are non-zero.

5.14 NH_3, HCl, CO_2, C_2H_4. The homonuclear diatomics cannot have vibrations which give a change in dipole moment. All of them can give lines in a Raman spectrum as they will all have polarisabilities that change during at least some vibrations.

5.15 $\varepsilon_v = (v + \tfrac{1}{2})\bar\omega_0 - (v + \tfrac{1}{2})^2\chi\bar\omega_0$
$\varepsilon_{v+1} = (v + \tfrac{3}{2})\bar\omega_0 - (v + \tfrac{3}{2})^2\chi\bar\omega_0$
$\Delta\varepsilon = \bar\omega_0 - 2(v + 1)\chi\bar\omega_0.$

χ is the order 10^{-2} so when v is of the order 10^2, ΔE becomes zero.

5.16 (i) $\varepsilon_1 - \varepsilon_0 = \bar\omega_0(1 - 2\chi)$ from result to question 5.15. Similarly, $\varepsilon_2 - \varepsilon_0 = 2\bar{\omega}_0$ $(1 - 3\chi)$. Eliminating $\bar{\omega}_0$ gives $\chi = 0.006$ (dimensionless).

(ii) Then,

$$\bar{\omega}_0 = 2169\,\text{cm}^{-1}.$$

Using

$$f_0 = \frac{1}{2\pi}\sqrt{\left(\frac{k}{\mu}\right)}$$

with $f_0 = \bar{\omega}_0 c$ gives
$$k = \mu(2\pi\bar{\omega}_0 c)^2.$$

But

$$\mu = \frac{1.993 \times 2.656}{1.993 + 2.656} \times 10^{-26}.$$

Substituting values gives $k = 1901\,\text{Nm}^{-1}$. The observed values of $\bar{\omega}_0$ and k are $2170.2\,\text{cm}^{-1}$ and $1902\,\text{Nm}^{-1}$ respectively.

5.17 (i) If $^1_1\text{H} \equiv m_1$ and $^{35}_{17}\text{Cl} \equiv m_2$ then

$$\mu_{\text{HCl}} = \frac{m_1 m_2}{m_1 + m_2}, \quad \mu_{\text{DCl}} = \frac{2m_1 m_2}{2m_1 + m_2}$$
$$= \frac{m_1 m_2}{m_1 + \frac{1}{2}m_2}.$$

Hence $\mu_{\text{HCl}} < \mu_{\text{DCl}}$.

(ii) Similarly,

$$\frac{f_0^{\text{HCl}}}{f_0^{\text{DCl}}} = \sqrt{\left[\frac{2(m_1 + m_2)}{2m_1 + m_2}\right]}.$$

If we ignore m_1 compared to m_2 then

$$f_0^{\text{HCl}} \approx \sqrt{2} f_0^{\text{DCl}}.$$

(iii)

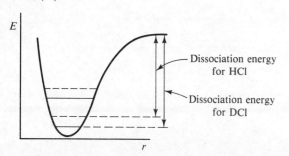

Dissociation energy for HCl

Dissociation energy for DCl

(iv) Isotopes cause multiple lines where, without isotopes, only one would occur.

5.18 The α-particles are believed to tunnel through the barrier owing to their wavefunctions penetrating the barrier.

5.19 As the overall product $\psi_2 \hat{\alpha} \psi_1$ must have even parity, we must have

odd × even × odd = even

or

even × even × even = even.

That is, unlike the case of electric dipole allowed transitions, Raman transition can only occur between states of the same parity. As a result, g↔g and u↔u transitions are Raman allowed in molecules with a centre of symmetry. These are not infrared allowed, and vice versa. Therefore the infrared and Raman transitions are mutually exclusive.

5.20 In order to convert from wavenumbers to frequency in hertz, multiply by the speed of light in cm s^{-1}. This gives a frequency of about $2.4 \times 10^{10}\,\text{Hz}$. Thus the inversion occurs about 10^{10} times every second.

5.21 For an electric dipole transition,

$$-e \int_{-\infty}^{\infty} \psi_1(x) x \psi_0(x)\,dx \neq 0.$$

Putting in the correct terms for the wavefunctions

$$\psi_0(x) = \left(\frac{\alpha}{\pi}\right)^{1/4} e^{-\alpha x^2/2}$$

$$\psi_1(x) = \frac{1}{2^{1/2}}\left(\frac{\alpha}{\pi}\right)^{1/4} e^{-\alpha x^2/2} 2\alpha^{1/2} x.$$

Thus the integral becomes

$$-e\alpha\left(\frac{2}{\pi}\right)^{1/2} \int_{-\infty}^{\infty} e^{-\alpha x^2} x^2\,dx$$

$$= -e\alpha\left(\frac{2}{\pi}\right)^{1/2} 2 \int_{0}^{\infty} e^{-\alpha x^2} x^2\,dx$$

by symmetry owing to $e^{-\alpha x^2} x^2$ being an even function symmetrical about $x = 0$.

Now the integral simplifies to give

$$-e\alpha\left(\frac{2}{\pi}\right)^{1/2} 2\frac{1}{4}\left(\frac{\pi}{\alpha^3}\right)^{1/2}$$

$$= -\frac{e}{(2\alpha)^{1/2}}.$$

The fact that $-e\int_{-\infty}^{\infty}\psi_2(x)x\psi_0(x)dx = 0$ follows from symmetry as the product $\psi_2(x)x\psi_0(x)$ is an odd function.

5.22

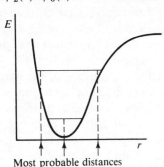

Most probable distances

(i), (ii) the most probable distances occur at the maxima in the probability density distributions. (iii) The vertical transition is from $v'' = 0$ to $v' = 0$.

(iv) There are two possibilities, from $v'' = 4$ to $v' = 3$ or $v' = 4$.

(v) From $v'' = 0$ the transition is most likely to be to $v' = 1$ because the molecule is more likely to be at that internuclear separation in $v' = 1$ than $v' = 0$. Look at fig. 5.20 to confirm this by examining the form of the probability density distributions. From $v'' = 4$, two things might happen. There could be a transition to $v' = 4$; but of more interest is that the other vertical transition from $v'' = 4$ goes into the continuum of the excited state. This means that at this separation in the excited state the atoms are too close together and the repulsive energy will cause them to fly apart. This transition therefore causes the molecule to dissociate.

Notice that the possible outcomes in a transition between two electronic states depends on the relative orientations of the potential energy curves.

5.23 The wavefunctions multiply and the energies add:

$$\psi(x, y) = \psi_v(x)\psi_v(y),$$

$$E_{v_x+v_y} = (v_x + v_y + 1)hf_0.$$

When $E_{v_x+v_y} = hf_0$ there is only one solution $v_x = v_y = 0$. For $2hf_0$, $v_x = 0$, $v_y = 1$ or vice

versa; thus this solution is doubly degenerate. Solutions for $3hf_0$ are four-fold degenerate, and for $4hf_0$ are five-fold degenerate.

5.24 The molecule can rotate clockwise or counter-clockwise while having the same energy.

5.25
$$\int_0^{2\pi} |\psi(\phi)|^2 d\phi = \int_0^{2\pi} A^2 \cos^2 m\phi\, d\phi$$

$$= A^2 \int_0^{2\pi} \tfrac{1}{2}(1 + \cos 2m\phi)d\phi$$

$$= \frac{A^2}{2}\left[\phi + \frac{1}{2m}\sin 2m\phi\right]_0^{2\pi}$$

$$= A^2\pi.$$

Thus we must have $A = 1/\sqrt{\pi}$.

5.26 The rotational energy depends on the reciprocal of the moment of inertia. In an obvious notation

$$E_{HCl} \propto \frac{1}{I_{HCl}}; \quad E_{DCl} \propto \frac{1}{I_{DCl}}.$$

But

$$I_{HCl} = \frac{m_H m_{Cl}}{m_H + m_{Cl}}r^2; \quad I_{DCl} = \frac{m_D m_{Cl}}{m_D + m_{Cl}}r^2.$$

Thus

$$\frac{E_{HCl}}{E_{DCl}} = \frac{(m_D + m_{Cl})m_H}{(m_H + m_{Cl})m_D}$$

$$\approx \tfrac{1}{2}$$

and the rotational energies of $_1^1H_{17}^{35}Cl$ are about half as large as those of $_1^1H_{17}^{35}Cl$.

5.27 (i) $E = (v + \tfrac{1}{2})hf_0 - (v + \tfrac{1}{2})^2\chi hf_0$
(ii) $E_J = BJ(J + 1) - DJ^2(J + 1)^2$
$E_T = E_v + E_J$
and
$\Delta E_T = E_{v+1} - E_v + E_{J+1} - E_J$
$= hf_0 - 2(v + 1)\chi hf_0 + 2B$
$+ 4D(J + 1)^3.$

5.28 The distribution is $n_1 = n_0 e^{-\Delta E/kT}$.
(i) From table 5.1 E for an electronic transition is of the order $10^2\, kJ\, mol^{-1}$. The value of k is $1.381 \times 10^{-23}\, J\, K^{-1}$, which becomes $1.382 \times 10^{-23} \times 6.022 \times 10^{23}$ $J\, K^{-1}\, mol^{-1}$. If we take T as $298\, K$ then n_1/n_0 takes the value of order 10^{-18}. That is, all the electrons in an atom

will be in the lowest energy level they can reach. Upper levels will only become occupied if energy is supplied to the atom.

(ii) A similar calculation for vibrational states shows that n_1/n_0 is of the order 10^{-2}; i.e. about 1% of molecules may be in an upper vibrational energy level at room temperature.

(iii) For rotational states the occupancy of upper levels can be significant e.g. of the order 60%.

(iv) Owing to the small difference in energy between nuclear spin states, the difference in population is very much less than 1%.

5.29 CH_3^+, H_2 and H^- would not give an e.s.r. spectrum because all their electrons are paired. SO_3^- would give only one line because the nuclei of sulphur and oxygen each have zero spin. Therefore there can be no hyperfine splitting.

5.30 Deuterium has a nuclear spin of 1. Thus it has three spin states corresponding to $m_I = +1, 0, -1$. Call these α, γ and β respectively. Then, for example, there are six ways of obtaining a total spin of α: $\alpha\alpha\beta$, $\alpha\beta\alpha$, $\alpha\gamma\gamma$, $\beta\alpha\alpha$, $\gamma\alpha\gamma$, $\gamma\gamma\alpha$. By summing the other combinations we find that the pattern is a weighting of $1:3:6:7:6:3:1$ for total spins of $3\alpha, 2\alpha, \alpha, 0, \beta, 2\beta, 3\beta$ respectively. The spectrum should consist of seven lines with intensities given by these ratios.

5.31 We would expect the six hydrogens to be equivalent, thus the spectrum should consist of seven lines in the ratio $1:6:15:20:15:6:1$. The problem is to explain how the hydrogens influence the π electron which has zero charge density at the hydrogen atoms and, indeed, zero charge density at the nuclei of the carbons. The spectrum actually does have the form we predict, the explanation being given by the theory of spin polarisation. More details can be found in the books listed in the bibliography.

5.32 The CH_2 group brings about a splitting in ratio $1:3:3:1$, i.e. a quartet. The CH_3 group causes a splitting in ratio $1:4:6:4:1$, i.e. a quintet. Each of the quartet lines will therefore be split into a quintet, giving 20 lines altogether.

5.33 (i) The charge densities are given by the formula of M.4.7. The results are 0.181, 0.065, 0.065, 0.181, 0.181, 0.065, 0.065, 0.181, 0.00, 0.00 for atoms 1 to 10 respectively. (ii) Thus atoms 1, 4, 5, 8 are equivalent in the time spent there by the electron, as are atoms 2, 3, 6, and 7. Atoms 9 and 10 are equivalent in so far as there is zero charge density for them. (iii) If we assume that the equivalence of the carbons runs to equivalence of the hydrogen atoms attracted to them, then the hydrogens at atoms 1, 4, 5, 8 and 2, 3, 6, 7 form two sets. (iv) We should expect there to be two sets of quintets, one splitting the other, i.e. 25 lines altogether. This is observed.

5.34 (i) We have

$$B_z = \frac{hf}{g_e\mu_B}$$

$$= \frac{6.626 \times 10^{-34} \times 9 \times 10^9}{2.0023 \times 9.274 \times 10^{-24}}$$

$$= 0.3211 \text{ T};$$

(ii) when $g = 2.000$, $B_z = 0.3215$ T.
In terms of gauss, these values are 3.211 kG and 3.215 kG respectively.

5.35 $\Delta E = 2.0023 \times 9.274 \times 10^{-24} \times 0.3$
$= 5.571 \times 10^{-24}$ J.

Thus

$$\frac{n_1}{n_0} = \exp\left[-\frac{5.571 \times 10^{-24}}{1.381 \times 10^{-23} \times 300}\right]$$

$$\approx 0.987.$$

e.s.r. must be a very sensitive technique to observe the absorption of energy when the difference in populations is so small.

5.36 (i) Antiparallel. $\mathbf{l}\cdot\mathbf{s} = |\mathbf{l}||\mathbf{s}|\cos\theta$, which is a minimum when $\theta = \pi$.

(ii) A fall.

(iii) A rise.

(iv)

The frequency will be lowered.

5.37 Ethanol is CH_3CH_2OH. The CH_3 proton resonance will be split into a triplet by the CH_2 protons. The CH_2 resonance will be split into a quartet by the CH_3 protons, and also each of these will be split into a doublet by the OH proton. The latter's resonance will be split into a triplet by the CH_2 protons. Thus we predict the pattern

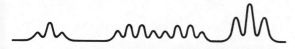

In reality the two CH_2 quartets overlap one another. If the ethanol is not pure but contains a trace of water, then the OH proton undergoes rapid exchange with the water. This has the effect of 'averaging out' the spin of the OH proton so that the effects of individual α or β states disappear. The OH resonance is then not split, nor does the OH proton split the CH_2 resonance.

5.38 There are two types of proton, corresponding to the two absorption regions. These are in the ratio 6:1. Thus there is a set of 12, and a set of 2 equivalent protons. None of them can be bonded to the oxygen. Thus either the compound is an aldehyde, ketone or ether. The larger peaks occur at a τ-value corresponding to methyl protons. A set of 12, suggests four CH_3 groups. Trial and error rules out an aldehyde or ketone. The structure is

$$
\begin{array}{ccc}
CH_3 & & CH_3 \\
\diagdown & & \diagup \\
& C - O - C & \\
\diagup \; \diagdown & & \diagup \; \diagdown \\
CH_3 \quad H & & H \quad CH_3
\end{array}
$$

5.39 (i) The abundance of $^{13}_{6}C$ is very low, about 1%. Such a low proportion means that the spectrometer has to be made very sensitive. Use has to be made of statistical sampling techniques.

(ii) In $CHCl_3$ the resonance will be split into a doublet. There are three spin states for a spin of 1 corresponding to $m_I = +1, 0, -1$. Thus in $CDCl_3$ the carbon resonance will be split into a triplet.

5.40 In the chair form cyclohexane there are two types of hydrogen; the axial and equatorial. At low temperatures one chair form predominates and resonances for both axial and equatorial protons show up. At higher temperatures one chair form flips over into the other:

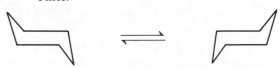

The axial and equatorial hydrogens swap roles and a similar situation exists as for ethanol above. The interconversion occurs so rapidly that only the average of the two signals survives and we observe a single resonance.

5.41 (i) $\hbar$.

(ii) Also $\hbar$.

(iii) Bosons.

Chapter 6

6.1 (i) $d^*d^* = e^2 + f^2$;

(ii) $\mathrm{Im}\,(ic) = \dfrac{1}{2i}(ic - i^*c^*)$

$$= \dfrac{1}{2i}(ic + ic^*) = \tfrac{1}{2}(c + c^*) = \mathrm{Re}\,c;$$

(iii) because $\cos\theta + i\sin\theta = e^i\theta$,

$$(\cos\theta + i\sin\theta)^n = (e^{i\theta})^n = e^{in\theta}$$
$$= \cos n\theta + i\sin n\theta.$$

This is known as de Moivre's theorem;

(iv) $\mathrm{Re}\,(cd^* + c^*d)$
$$= \tfrac{1}{2}[(cd^* + c^*d) + (cd^* + c^*d)^*]$$
$$= \tfrac{1}{2}[(cd^* + c^*d) + (c^*d + cd^*)]$$
$$= cd^* + c^*d$$
$$= 2\,\mathrm{Re}\,(cd^*) = 2(ae + bf);$$

(v) $\mathrm{Re}\,(cd^* - c^*d) = \mathrm{Re}\,(2i\,\mathrm{Im}\,cd^*)$
$$= \mathrm{Re}\,(2ae + 2bf) = 2(ae + bf).$$

6.2 (i) $\displaystyle\int_0^{2\pi} 1\,d\phi = 2\pi;$

(ii) $\displaystyle\int_0^{2\pi} e^{in\phi}\,d\phi = \left[\dfrac{1}{in}e^{in\phi}\right]_0^{2\pi}$

$$= \dfrac{1}{in}\left[\cos n\phi + i\sin n\phi\right]_0^{2\pi}$$

$= 0$ because $\cos n2\pi = \cos 0 = 1$ and $\sin n2\pi = \sin 0 = 0$.

6.3 (i) $-\hbar^2\dfrac{\partial^2(A\mathrm{e}^{\mathrm{i}m\phi})}{\partial\phi^2} = -\hbar^2\mathrm{i}m\dfrac{\partial(A\mathrm{e}^{\mathrm{i}m\phi})}{\partial\phi}$

$$= \hbar^2 m^2 A\mathrm{e}^{\mathrm{i}m\phi}.$$

Thus comparing with $EA\mathrm{e}^{\mathrm{i}m\phi}$, shows

$$E = m^2\hbar^2.$$

(ii) $\displaystyle\int_0^{2\pi}\psi_m^*(\phi)\psi_m(\phi)\,\mathrm{d}\phi$

$$= \int_0^{2\pi} A\mathrm{e}^{-\mathrm{i}m\phi}A\mathrm{e}^{\mathrm{i}m\phi}\,\mathrm{d}\phi$$

$$= A^2\int_0^{2\pi}\mathrm{d}\phi$$

$$= A^2 2\pi,$$

which means that $A^2 = 1/2\pi$, so $A = \pm 1/\sqrt{(2\pi)}$.

6.4 (i) $|\mathbf{r}|^2 = 2^2 + 4^2 + 6^2 = 56$.
Thus, $\mathbf{r}$ is normalised by putting $(1/\sqrt{56})(2\mathbf{i} + 4\mathbf{j} + 6\mathbf{k})$.

(ii) There are an infinite number of vectors orthogonal to $\mathbf{r}$. You may think of a cylinder with $\mathbf{r}$ as its long axis. All the possible radii of the cylinder will be orthogonal to $\mathbf{r}$. If we write an orthogonal vector as $\mathbf{r}_1 = c_1\mathbf{i} + c_2\mathbf{j} + c_3\mathbf{k}$ then $\mathbf{r}\cdot\mathbf{r}_1 = 2c_1 + 4c_2 + 6c_3 = 0$.
We could choose $c_1 = 4$, $c_2 = -2$ and $c_3 = 0$ to form $\mathbf{r}_1 = 4\mathbf{i} - 2\mathbf{j}$ as one solution.

6.5 Yes. We have $\mathbf{r}_1\cdot\mathbf{r}_2 = 3\times 45 + 7\times 0 - 5\times 27 = 0$.

6.6 (i) $\mathbf{V}_1^*\cdot\mathbf{V}_1 = (-2\mathrm{i})(2\mathrm{i})$
$\qquad\qquad + 3\times 3 + 4\times 4 + (-5\mathrm{i})(5\mathrm{i})$
Compare $\mathbf{V}^*\cdot\mathbf{V} = c_1^*c_1 + c_2^*c_2 + \cdots + c_n^*c_n$.

(ii) $|\mathbf{V}_1| = \sqrt{54}$.

(iii) Normalised, $\mathbf{V}_1 = (1/\sqrt{54})(2\mathrm{i}\mathbf{v}_1 + 3\mathbf{v}_2 + 4\mathbf{v}_3 + 5\mathrm{i}\mathbf{v}_4)$.

(iv) $\mathbf{V}_1^*\cdot\mathbf{V}_2 = (2\mathrm{i}\mathbf{v}_1 + 3\mathbf{v}_2 + 4\mathbf{v}_3 + 5\mathrm{i}\mathbf{v}_4)^*$
$\qquad\qquad \cdot(6\mathrm{i}\mathbf{v}_1 + 2\mathbf{v}_2 + 3\mathbf{v}_3 - 6\mathrm{i}\mathbf{v}_4)$
$\qquad = (-2\mathrm{i})(6\mathrm{i}) + 3\times 2 + 4\times 3$
$\qquad\qquad + (-5\mathrm{i})(-6\mathrm{i})$
$\qquad = 12 + 6 + 12 - 30$
$\qquad = 0.$

Hence $\mathbf{V}_1$ and $\mathbf{V}_2$ are orthogonal.

6.7 (i) $\dfrac{\mathrm{d}(x^2)}{\mathrm{d}x} = 2x;\qquad \dfrac{\mathrm{d}(\mathrm{e}^{kx})}{\mathrm{d}x} = k\mathrm{e}^{kx};$
$\dfrac{\mathrm{d}(\mathrm{e}^{\mathrm{i}kx})}{\mathrm{d}x} = \mathrm{i}k\mathrm{e}^{\mathrm{i}kx}.$

(ii) $\dfrac{\mathrm{d}^2(x^2)}{\mathrm{d}x^2} = 2;\qquad \dfrac{\mathrm{d}^2(\mathrm{e}^{kx})}{\mathrm{d}x^2} = k^2\mathrm{e}^{kx};$
$\dfrac{\mathrm{d}^2(\mathrm{e}^{\mathrm{i}kx})}{\mathrm{d}x^2} = -k^2\mathrm{e}^{\mathrm{i}kx}.$

They are all eigenfunctions except x^2. The eigenvalues are as shown.

6.8 $\left(x\dfrac{\mathrm{d}}{\mathrm{d}x} - \dfrac{\mathrm{d}}{\mathrm{d}x}x\right)f(x)$

$$= x\dfrac{\mathrm{d}}{\mathrm{d}x}f(x) - \dfrac{\mathrm{d}}{\mathrm{d}x}xf(x)$$

$$= xf'(x) - [f(x) + xf'(x)]$$

$$= -f(x).$$

6.9 $\left[x, \dfrac{\mathrm{d}^2}{\mathrm{d}x^2}\right]f(x)$

$$= x\dfrac{\mathrm{d}^2}{\mathrm{d}x^2}f(x) - \dfrac{\mathrm{d}^2}{\mathrm{d}x^2}xf(x)$$

$$= xf''(x) - (\mathrm{d}/\mathrm{d}x)[f(x) + xf'(x)]$$

$$= xf''(x) - [f'(x) + f'(x) + xf''(x)]$$

$$= -2f'(x)\text{ or } -2(\mathrm{d}/\mathrm{d}x)f(x),$$

i.e.

$$\left[x, \dfrac{\mathrm{d}^2}{\mathrm{d}x^2}\right] = -2\dfrac{\mathrm{d}}{\mathrm{d}x},$$

so they do not commute.

6.10 (i) $[\hat{L}_x, y]F = \dfrac{\hbar}{\mathrm{i}}\left[\left(y\dfrac{\partial}{\partial z} - z\dfrac{\partial}{\partial y}\right)yF\right.$

$$\left. - y\left(y\dfrac{\partial}{\partial z} - z\dfrac{\partial}{\partial y}\right)F\right]$$

$$= \dfrac{\hbar}{\mathrm{i}}\left[\left(y\dfrac{\partial y}{\partial z} - z\right)F\right.$$

$$+ y\left(y\dfrac{\partial F}{\partial z} - z\dfrac{\partial F}{\partial y}\right)$$

$$\left. - y\left(y\dfrac{\partial F}{\partial z} - z\dfrac{\partial F}{\partial y}\right)\right]$$

$$= \mathrm{i}\hbar zF\text{ because }\partial y/\partial z = 0$$

so
$$[\hat{L}_x, y] = i\hbar z;$$

(ii) $[\hat{L}_x\hat{L}_y - \hat{L}_y\hat{L}_x]F$

$$= -\hbar^2\left[\left(y\frac{\partial}{\partial z} - z\frac{\partial}{\partial y}\right)\right.$$

$$\cdot\left(z\frac{\partial}{\partial x} - x\frac{\partial}{\partial z}\right)$$

$$-\left(z\frac{\partial}{\partial x} - x\frac{\partial}{\partial z}\right)$$

$$\left.\cdot\left(y\frac{\partial}{\partial z} - z\frac{\partial}{\partial y}\right)\right]F$$

$$= -\hbar^2\left[\left(y\frac{\partial}{\partial z} - z\frac{\partial}{\partial y}\right)\right.$$

$$\cdot\left(z\frac{\partial F}{\partial x} - x\frac{\partial F}{\partial z}\right)$$

$$-\left(z\frac{\partial}{\partial x} - x\frac{\partial}{\partial z}\right)$$

$$\left.\cdot\left(y\frac{\partial F}{\partial z} - z\frac{\partial F}{\partial y}\right)\right]$$

$$= -\hbar^2\left[\left(y\frac{\partial F}{\partial x} + yz\frac{\partial^2 F}{\partial x\partial z} - xy\frac{\partial^2 F}{\partial z^2}\right)\right.$$

$$-\left(z^2\frac{\partial F}{\partial x\partial y} - xz\frac{\partial^2 F}{\partial y\partial z}\right)$$

$$-\left(yz\frac{\partial^2 F}{\partial x\partial z} - z^2\frac{\partial^2 F}{\partial x\partial y}\right)$$

$$\left.+\left(xy\frac{\partial^2 F}{\partial z^2} - x\frac{\partial F}{\partial y} - xz\frac{\partial^2 F}{\partial y\partial z}\right)\right]$$

$$= -\hbar^2\left[y\frac{\partial F}{\partial x} - x\frac{\partial F}{\partial y}\right]$$

$$= \hbar^2\left[x\frac{\partial F}{\partial y} - y\frac{\partial F}{\partial x}\right]$$

$$= i\hbar\hat{L}_z F,$$

which proves the result.

(iii) (iv), (v) all these results follow, like those for (i) and (ii) provided you are careful. The cyclic nature of the relationship between $\hat{L}_x, \hat{L}_y, \hat{L}_z$ suggests that $[\hat{L}_x, \hat{L}_y] = i\hbar\hat{L}_z$; $[\hat{L}_y, \hat{L}_z] = i\hbar\hat{L}_x$; $[\hat{L}_z, \hat{L}_x] = i\hbar\hat{L}_y$.

6.11 $[\hat{p}_x, x]F = \dfrac{\hbar}{i}\left[\dfrac{\partial}{\partial x}x - x\dfrac{\partial}{\partial x}\right]F$

$$= \frac{\hbar}{i}\left[F + x\frac{\partial F}{\partial x} - x\frac{\partial F}{\partial x}\right]$$

$$= \frac{\hbar}{i}F$$

i.e.

$$[\hat{p}_x, x] = \frac{\hbar}{i}.$$

6.12 The state vector is complex and a vector. On both counts it cannot correspond to observables, which have to be real numbers.

6.13 (i) $\langle S|\hat{H}\hat{p}|S\rangle = \langle S|\hat{H}p|S\rangle = p\langle S|\hat{H}|S\rangle$
$$= p\langle S|E|S\rangle$$
so
$$\langle S|\hat{H}\hat{p}|S\rangle = pE.$$

(ii) $\langle S|\hat{p}\hat{H}|S\rangle = Ep$ by a similar argument.

(iii) Therefore $\langle S|\hat{H}\hat{p}|S\rangle - \langle S|\hat{p}\hat{H}|S\rangle = 0$ because $pE = Ep$. (These are real numbers, not operators.) Thus $[\hat{H}, \hat{p}] = 0$. Therefore they do commute.

There is an important theorem about this type of behaviour. See section 6.9.

6.14 We found on p. 114 that
$$|\mathbf{V}|^2 = \mathbf{V}^*\cdot\mathbf{V} = \sum_i c_i^* c_i.$$
Similarly,
$$\langle S|\hat{H}|S\rangle = \left(\sum_j c_j^*\langle N_j|\right)\left(\sum_i c_i E_i|N_i\rangle\right)$$
$$= \sum_i\sum_j c_j^* c_i E_i\langle N_j|N_i\rangle$$
$$= \sum_i\sum_j c_j^* c_i E_i\delta_{i,j}$$
$$= \sum_i c_i^* c_i E_i.$$

6.15 (i) The total probability must be unity. Therefore,
$$c_1^* c_1 + c_2^* c_2 = 1.$$

(ii) $\langle S|S\rangle = \left(\sum_j c_j^*\langle N_j|\right)\left(\sum_i c_i|N_i\rangle\right)$
$$= \sum_i\sum_j c_j^* c_i\langle N_j|N_i\rangle$$
$$= \sum_i\sum_j c_j^* c_i\delta_{i,j}$$
$$= \sum_i c_i^* c_i.$$

Thus

$$\sum_i c_i{}^*c_i = 1 \text{ if } \langle S|S\rangle = 1.$$

(iii) Normalisation guarantees that the squares, $c_i{}^*c_i$, of the expansion coefficients can serve as measures of probabilities.

6.16 With $|S'\rangle = e^{i\phi}|S\rangle$ we have $\langle S'| = \langle S|e^{-i\phi}$.

Thus

$$\begin{aligned}\langle S'|\hat{H}|S'\rangle &= \langle S|e^{i\phi}\hat{H}e^{i\phi}|S\rangle\\ &= e^{-i\phi}e^{i\phi}\langle S|\hat{H}|S\rangle\\ &= \langle S|\hat{H}|S\rangle\\ &= E.\end{aligned}$$

This proves that we only know the precise form of an eigenstate to within an arbitrary factor – the phase factor – $e^{i\phi}$. Once again, remember that, like a wavefunction, it is not the state vector itself, but its square, that is important.

6.17

(i), (ii)
$$\begin{aligned}x_{ex} &= \int_0^l \psi_1(x)x\psi_1(x)dx\\ &= \frac{2}{l}\int_0^l x\sin^2\frac{\pi x}{l}dx\\ &= \frac{2}{l}\left[\frac{x}{2}\left(\frac{x}{2} - \frac{1}{2k}\sin 2kx\right)\right.\\ &\qquad\left. - \frac{1}{2k^2}\cos 2kx\right]_0^l\end{aligned}$$

where $k = \pi/l$.

Thus, $x_{ex} = l/2$. This is what we would expect from classical theory.

6.18
$$\begin{aligned}p_{ex} &= \int_0^l \psi_1{}^*(x)\hat{p}_x\psi_1(x)dx\\ &= \frac{h}{\pi l\mathrm{i}}\int_0^l \sin\frac{\pi x}{l}\frac{\mathrm{d}}{\mathrm{d}x}\sin\frac{\pi x}{l}dx\\ &= \frac{h}{l^2\mathrm{i}}\int_0^l \sin\frac{\pi x}{2}\cos\frac{\pi x}{l}dx\\ &= \frac{h}{2l^2\mathrm{i}}\int_0^l \sin\frac{2\pi x}{l}dx\\ &= -\frac{h}{4\pi l\mathrm{i}}\left[\cos\frac{2\pi x}{l}\right]_0^l\\ &= 0.\end{aligned}$$

This reflects our intuitive idea that the particle is to be found moving to the left as often as to the right. Hence the average momentum should be zero.

6.19 (i) First note that

$$\begin{aligned}\mathcal{T}\psi_{1s} &= -\frac{1}{2\sqrt{\pi}}\frac{1}{r^2}\frac{\partial}{\partial r}\left(r^2\frac{\partial e^{-r}}{\partial r}\right)\\ &= -\frac{1}{2\sqrt{\pi}}\frac{1}{r^2}\frac{\partial}{\partial r}(-r^2 e^{-r})\\ &= -\frac{1}{2\sqrt{\pi}}\frac{1}{r^2}(-2re^{-r} + r^2 e^{-r}).\end{aligned}$$

Then

$$\begin{aligned}\int\psi_{1s}\mathcal{T}\psi_{1s}d\tau &= \frac{1}{2\pi}\int_0^\infty (2re^{-2r}\\ &\qquad - r^2 e^{-2r})dr\\ &\qquad \times \int_0^\pi \sin\theta d\theta \int_0^{2\pi} d\phi\\ &= \frac{1}{2\pi}\left[\tfrac{1}{2} - \tfrac{1}{4}\right](2)(2\pi)\\ &= +\tfrac{1}{2}\end{aligned}$$

(ii)
$$\begin{aligned}\int\psi_{1s}\hat{V}\psi_{1s}d\tau &= -\frac{1}{\pi}\int_0^\infty re^{-2r}dr\\ &\qquad \times \int_0^\pi \sin\theta d\theta \int_0^{2\pi} d\phi\\ &= -1.\end{aligned}$$

(iii) The virial theorem should be that $2\mathcal{T}_{ex} = -V_{ex}$. It is obeyed.

6.20 (i)
$$\begin{aligned}[\hat{H}_t, t]F &= i\hbar\left[\frac{\partial}{\partial t}t - t\frac{\partial}{\partial t}\right]F\\ &= i\hbar F.\end{aligned}$$

Therefore, $\hat{C} \equiv \hbar$.

(ii) $\Delta E\Delta t \geqslant \hbar/2$.

(iii) Using $E = hf$,

$$\Delta f\Delta t \geqslant 1/4\pi.$$

(iv) $\Delta f = 2.65\,\text{MHz}$ and $7.96\,\text{MHz}$.

(v) The ratio is of the order 10^8. States with long life times give sharp spectral lines, and vice versa.

Chapter 7

7.1 (i) $\hat{S}_+|\alpha\rangle = 0$;   (ii) $\hat{S}_+|\beta\rangle = N_+|\alpha\rangle$;
(iii) $\hat{S}_-|\alpha\rangle = N_-|\beta\rangle$;   (iv) $\hat{S}_-|\beta\rangle = 0$

7.2 (i) For example, $\hat{L}_z|x\rangle = (1/\sqrt{2})(\hbar|1,1\rangle$
$- \hbar|1, -1\rangle) \neq$ a number
$\times (1/\sqrt{2})(|1,1\rangle + |1, -1\rangle)$ but

$$\hat{L}_z \frac{1}{\sqrt{2}}(|x\rangle + i|y\rangle)$$
$$= \tfrac{1}{2}\hat{L}_z[(|1,1\rangle + |1,-1\rangle) + (|1,1\rangle - |1,-1\rangle)]$$
$$= \tfrac{1}{2}\hbar[(|1,1\rangle - |1,-1\rangle) + (|1,1\rangle + |1,-1\rangle)]$$
$$= \hbar\frac{1}{\sqrt{2}}[i|y\rangle + |x\rangle].$$

So the eigenvalue is $\hbar$. It is $-\hbar$ for
$(1/\sqrt{2})(|x\rangle - i|y\rangle)$.

(ii) Note that $(1/\sqrt{2})(|x\rangle + i|y\rangle) = |1,1\rangle$
and that $(1/\sqrt{2})(|x\rangle - i|y\rangle) = |1, -1\rangle$.
Then $\hat{L}_+$ acting on the former combination and $\hat{L}_-$ acting on the latter have no effect: it is impossible to go off the top or bottom of the ladder $\hat{L}_+$ acting on $(1/\sqrt{2})(|x\rangle - i|y\rangle)$ and $\hat{L}_-$ acting on $(1/\sqrt{2})(|x\rangle + i|y\rangle)$ both result in the state $|z\rangle$ or $|1,0\rangle$ owing to their effects as ladder operators.

7.3 (i) $\hat{H}|l,m\rangle = l(l+1)\hbar|l,m\rangle$.
Therefore, $E_l = l(l+1)\hbar^2/2I$.
(ii) $B = \hbar^2/2I$.
(iii) $\hat{d}_z$.

7.4 (i) Spin–orbit coupling. See the diagram for the other answers.

7.5 (i) The transition dipole moment is proportional to

$$(\langle\alpha, \beta|\cos\theta + \langle\beta, \alpha|\sin\theta)|\hat{\mathbf{I}}_1 + \hat{\mathbf{I}}_2|\alpha, \alpha\rangle. \quad \text{(D)}$$

Now,
$$\hat{I}_{1x} + \hat{I}_{2x} = \tfrac{1}{2}(\hat{I}_{1+} + \hat{I}_{1-}) + \tfrac{1}{2}(\hat{I}_{2+} + \hat{I}_{2-})$$

so
$$(\hat{I}_{1x} + \hat{I}_{2x})|\alpha, \alpha\rangle = \tfrac{1}{2}N_-|\beta, \alpha\rangle$$
$$+ \tfrac{1}{2}N_-|\alpha, \beta\rangle.$$

Similarly,
$$(\hat{I}_{1y} + \hat{I}_{2y})|\alpha, \alpha\rangle = -\frac{N_-}{2i}|\beta, \alpha\rangle$$
$$-\frac{N_-}{2i}|\alpha, \beta\rangle$$

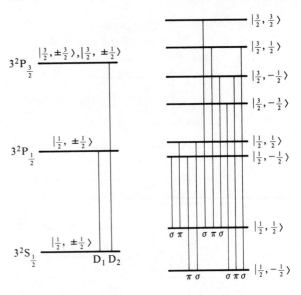

$3^2P_{\frac{3}{2}}$ $|\tfrac{3}{2}, \pm\tfrac{3}{2}\rangle, |\tfrac{3}{2}, \pm\tfrac{1}{2}\rangle$

$3^2P_{\frac{1}{2}}$ $|\tfrac{1}{2}, \pm\tfrac{1}{2}\rangle$

$3^2S_{\frac{1}{2}}$ $|\tfrac{1}{2}, \pm\tfrac{1}{2}\rangle$

$D_1 D_2$

$\sigma\,\pi$ $\sigma\,\pi\,\sigma$

$\pi\,\sigma$ $\sigma\,\pi\,\sigma$

$|\tfrac{3}{2}, \tfrac{3}{2}\rangle$
$|\tfrac{3}{2}, \tfrac{1}{2}\rangle$
$|\tfrac{3}{2}, -\tfrac{1}{2}\rangle$
$|\tfrac{3}{2}, -\tfrac{3}{2}\rangle$
$|\tfrac{1}{2}, \tfrac{1}{2}\rangle$
$|\tfrac{1}{2}, -\tfrac{1}{2}\rangle$
$|\tfrac{1}{2}, \tfrac{1}{2}\rangle$
$|\tfrac{1}{2}, -\tfrac{1}{2}\rangle$

Field off
Two lines in the spectrum
(the sodium D-lines)

Field on
Ten lines in the spectrum

The anomalous Zeeman effect of the sodium D-lines

and
$$(\hat{I}_{1z} + \hat{I}_{2z})|\alpha, \alpha\rangle = \hbar|\alpha, \alpha\rangle.$$

Therefore (D) becomes

$$\tfrac{1}{2}N_-\cos\theta - \frac{N_-}{2i}\cos\theta + \tfrac{1}{2}N_-\sin\theta$$
$$-\frac{N_-}{2i}\sin\theta$$
$$= \frac{N_-}{2}(1 + i)(\cos\theta + \sin\theta).$$

Thus the square is $(N_-/2)^2 (1 + i)^* (\cos\theta + \sin\theta)^* (1 + i)(\cos\theta + \sin\theta)$

$$= \left(\frac{N_-}{2}\right)^2 2(\cos^2\theta + 2\sin\theta\cos\theta + \sin^2\theta)$$
$$= \frac{N_-^2}{2}(1 + 2\sin\theta\cos\theta)$$
$$= \frac{N_-^2}{2}(1 + \sin 2\theta)$$

i.e. intensity $\propto (1 + \sin 2\theta)$.

(ii) So that they are orthogonal.
(iii) $\sin\theta|\alpha,\beta\rangle - \cos\theta|\beta,\alpha\rangle$ and $|\beta,\beta\rangle$
(iv)

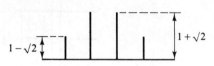

7.6

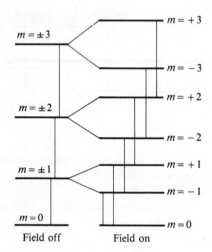

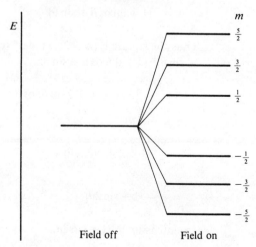

7.7 (i) m-values are $\frac{5}{2}, \frac{3}{2}, \frac{1}{2}, -\frac{1}{2}, -\frac{3}{2}, -\frac{5}{2}$.

(ii) $\langle j,m|\hat{H}_1|j,m\rangle = (ge\hbar/2m_e)mB_z$.
Thus the energies are, $\pm\frac{5}{2}, \pm\frac{3}{2}$, and $\pm\frac{1}{2}$
units of $(e\hbar/2m_e)gB_z$.

7.8 (i) $E_1 = \dfrac{\pi h^2}{32ml^3}\displaystyle\int_0^l \sin^2\dfrac{\pi x}{l}\sin\dfrac{\pi x}{l}\,dx$

$= \dfrac{\pi h^2}{128ml^3}\displaystyle\int_0^l \left[\sin(2n-1)\dfrac{\pi x}{l}\right.$

$\left. -\sin(2n+1)\dfrac{\pi x}{l} + 2\sin\dfrac{\pi x}{l}\right]dx$

$= \dfrac{\pi h^2}{128ml^2}\left[-\dfrac{1}{(2n-1)}\dfrac{l}{\pi}\cos(2n-1)\dfrac{\pi x}{l}\right.$

$+ \dfrac{1}{(2n+1)}\dfrac{l}{\pi}\cos(2n+1)\dfrac{\pi x}{l}$

$\left. -2\dfrac{l}{\pi}\cos\dfrac{\pi x}{l}\right]_0^l$

$= \dfrac{h^2}{64ml^2}\left[\dfrac{1}{2n-1}-\dfrac{1}{2n+1}+2\right].$

(ii) because $2n\pm 1$ is always odd and $\cos\theta$ is -1 at $(2n\pm 1)\pi$, i.e.

$$E_1 = \dfrac{h^2}{8ml^2}\dfrac{n^2}{4n^2-1}.$$

We know that the unperturbed energies are $n^2h^2/8ml^2$. Therefore

$$E_n = \dfrac{n^2h^2}{8ml^2}\left(1+\dfrac{1}{4n^2-1}\right).$$

(iii) $n=1$ gives a percentage change of 25%, $n=5$ gives $10^{-3}\%$ change.

7.9 The steps omitted in the proof of the variation theorem were:

$\langle T|\hat{H}-E_0|T\rangle = \sum_i\sum_j c_j^* c_i\langle N_j|\hat{H}|N_i\rangle$

$-E_0\sum_i\sum_j c_j^* c_i\langle N_j|N_i\rangle$

$= \sum_i\sum_j c_j^* c_i(E_i-E_0)\langle N_j|N_i\rangle$

$= \sum_i c_i^* c_i(E_i-E_0)$

because $\langle N_j|N_i\rangle = \delta_{i,j}$.
Finally, with

$\langle T|T\rangle = 1$, $\langle T|\hat{H}-E_0|T\rangle = \langle T|\hat{H}|T\rangle -E_0.$

(i) $\displaystyle\int_{-\infty}^{\infty}\psi_T^*\psi_T\,dx = 2A^2\int_0^{\infty}e^{-2Cx^2}dx$

$= 2A^2\dfrac{1}{2}\sqrt{\left(\dfrac{\pi}{2C}\right)}.$

Hence

$$A = \pm \left(\frac{2C}{\pi}\right)^{1/4}.$$

(ii) $\hat{H} = -\frac{\hbar^2}{2m}\frac{d^2}{dx^2} + \frac{1}{2}kx^2.$

(iii) $E_T = \frac{\hbar^2 A^2}{m}\int_0^\infty (-2Ce^{-2Cx^2}$

$$+ C^2 x^2 e^{-2Cx^2})dx$$

$$+ A^2 k \int_0^\infty x^2 e^{-2Cx^2}dx$$

$$= -\frac{\hbar^2 A^2}{m}\left[-2C\frac{1}{2}\sqrt{\left(\frac{\pi}{2C}\right)}\right.$$

$$\left. + 4C^2\frac{1}{4}\sqrt{\left(\frac{\pi}{8C^3}\right)}\right]$$

$$+ A^2 k \frac{1}{4}\sqrt{\left(\frac{\pi}{8C^3}\right)}.$$

After simplifying this becomes

$$E_T = \frac{\hbar^2}{2m}C + \frac{1}{8}kC^{-1}.$$

(iv) $\frac{\partial E_T}{\partial C} = \frac{\hbar^2}{2m} - \frac{1}{8}kC^{-2}$

$$= 0 \text{ when } C = (\pi/h)\sqrt{(mk)}.$$

(v) With this value, $E_T = \frac{1}{4}\frac{h}{\pi}\sqrt{\left(\frac{k}{m}\right)} = \frac{1}{2}hf_0.$

We have obtained the exact, or true, zero point energy. This was purely a

result of choosing a good trial wave-function. If, for example, we had chosen $\psi_T = Ax^c$ we would not have fared so well. This is the drawback with the variation theorem – much depends on how well we choose ψ_T.

7.10 (i) Because the molecule is homopolar, $\int \phi_A^* \hat{H}\phi_A dv = \int \phi_A^* \hat{H}\phi_B dv = $ the Coulomb integral, α.
Similarly the integrals involving cross terms e.g. $\int \phi_A^* \hat{H}\phi_B dv$ are equal to the same bond integral, β. S is the overlap integral, $\int \phi_A^* \phi_B dv$. We assume that ϕ_A and ϕ_B are normalised so, for example, $\int \phi_A^* \phi_A dv = 1$.

(ii) $\frac{\partial E_T}{\partial c_B}(c_A^2 + c_B^2 + 2c_A c_B S)$
$$+ E_T(2c_B + 2c_A S)$$
$$= 2c_A\beta + 2c_B\alpha.$$

(iii) Is trivial.

(iv) Expanding the determinant,

$$(\alpha - E_T)^2 - (\beta - E_T S)^2 = 0$$
$$(\alpha - E_T)^2 = (\beta - E_T S)^2$$

so

$$\alpha - E_T = +(\beta - E_T S)$$

or

$$\alpha - E_T = -(\beta - E_T S)$$

which gives the results.

(v) Multiplying the first of the simultaneous equations by c_2, the second by c_1, and comparing quickly gives $c_1^2 = c_2^2$. Hence $c_1 = \pm c_2$.

(vi) This is the same method as M.4.3.

APPENDIX A

Standard integrals

$$\int_0^\infty e^{-kx^2}dx = \frac{1}{2}\sqrt{\left(\frac{\pi}{k}\right)}$$

$$\int_0^\infty xe^{-kx^2}dx = \frac{1}{2k}$$

$$\int_0^\infty x^2 e^{-kx^2}dx = \frac{1}{4}\sqrt{\left(\frac{\pi}{k^3}\right)}$$

$$\int_0^\infty x^{2n} e^{-kx^2}dx$$

$$= \frac{1 \times 3 \times 5 \cdots \times (2n-1)}{2^{n+1}}\sqrt{\left(\frac{\pi}{k^{2n+1}}\right)}$$

$$\int_0^\infty x^{2n+1} e^{-kx^2}dx = \frac{n!}{2k^{n+1}}$$

$$\int_0^\infty x^n e^{-kx}dx = \frac{n!}{k^{n+1}} \quad (n > -1; \ k > 0)$$

$F(x)$	$\int F(x)dx$
e^{kx}	$\dfrac{1}{k}e^x$
$\sin kx$	$-\dfrac{1}{k}\cos kx$
$\cos kx$	$\dfrac{1}{k}\sin kx$
$\sin^2 kx$	$\dfrac{1}{2}\left(x - \dfrac{1}{2k}\sin 2kx\right)$
$\cos^2 kx$	$\dfrac{1}{2}\left(x + \dfrac{1}{2k}\sin 2kx\right)$
$\sin kx \cos^2 kx$	$-\dfrac{1}{3k}\cos^3 kx$
$\cos kx \sin^2 kx$	$\dfrac{1}{3k}\sin^3 kx$
$x\sin^2 kx$	$\dfrac{x}{2}\left(\dfrac{x}{2} - \dfrac{1}{2k}\sin 2kx\right) - \dfrac{1}{2k^2}\cos kx$

APPENDIX B

Some universal constants

Quantity	Symbol	Value and units
Avogadro's number	L	$6.022 \times 10^{23}\,\mathrm{mol}^{-1}$
Bohr magneton	β	$9.274 \times 10^{-24}\,\mathrm{J\,T}^{-1}$
Bohr radius	a_0	$5.292 \times 10^{-11}\,\mathrm{m}$
Boltzmann's constant	k	$1.381 \times 10^{-23}\,\mathrm{J\,K}^{-1}$
Electron mass	m_e	$9.109 \times 10^{-31}\,\mathrm{kg}$
Electron charge	$-e$	$1.602 \times 10^{-19}\,\mathrm{C}$
Planck's constant	h	$6.626 \times 10^{-34}\,\mathrm{J\,s}$
Proton mass	m_p	$1.673 \times 10^{-27}\,\mathrm{kg}$
Permittivity of free space	ε_0	$8.854 \times 10^{-12}\,\mathrm{C^2\,N^{-1}\,m^{-2}}$ or $\mathrm{F\,m}^{-1}$
Speed of light	c	$2.998 \times 10^{8}\,\mathrm{m\,s}^{-1}$

APPENDIX C

Greek letters used in the text

Symbol	Name
α	alpha
β	beta
γ	gamma
Δ, δ	delta
ε	epsilon
ζ	zeta
Θ, θ	theta
Λ, λ	lambda
μ	mu
ν	nu
Π, π	pi
ρ	rho
Σ, σ	sigma
τ	tau
Φ, ϕ	phi
Ψ, ψ	psi
Ω, ω	omega

APPENDIX D

Bibliography

There is a vast literature on quantum chemistry. The following seven books provide good, and more advanced, accounts of the main themes we have considered.

P.W. Atkins, *Molecular Quantum Mechanics* (2nd edn), Clarendon Press, Oxford, 1970.
—, *Quanta – a handbook of concepts*, Clarendon Press, Oxford, 1970.
C.A. Coulson, *Valence* (2nd edn), Oxford University Press, Oxford, 1961.
H. Eyring, J. Walter and G.E. Kimball, *Quantum Chemistry*, John Wiley, New York, 1944.
D.C. Harris and M.D. Bertolucci, *Symmetry and Spectroscopy*, Oxford University Press, New York, 1978.
R. McWeeny, *Coulson's Valence* (3rd edn), Oxford University Press, Oxford, 1979. (This is a revision of the superb book by Coulson cited above.)
L. Pauling and E.B. Wilson, *Introduction to Quantum Mechanics*, McGraw-Hill, New York, 1935.

References to original papers:

Electron spin
Goudsmit and Uhlenbeck published their work in:
G. Goudsmit and S. Uhlenbeck, *Naturwissenschaften* **13**, 953 (1925).

Heisenberg's uncertainty relation
This was published in:

W. Heisenberg, *Zeitschrift Für Physik* **43**, 172 (1927)

Planck's work and the old quantum theory
Planck's original work on black body radiation can be found in translation in:
H. Kangro (Ed.), *Classic Papers in Physics*, Vol. 1, Taylor & Francis, London, 1972.

For the original work on the photoelectric effect see:
H. Hertz, *Annalen der Physik*, **31**, 982 (1887).
P. Lenard, *Annalen der Physik* **2**, 359 (1900).
A. Einstein, *Annalen der Physik* **17**, 132 (1905).

The Geiger and Marsden experiments are in
H. Geiger and E. Marsden, *Phil. Mag.* **25**, 604 (1913).
Rutherford's original calculations are to be found in:
E. Rutherford, *Phil. Mag.* **21**, 669 (1911).
Balmer's paper can be found in translation in:
J.B. Marion, *A Universe of Physics*, John Wiley, New York, 1970.

Bohr's epoch making paper was:
N. Bohr, *Phil. Mag.*, **26**, 1 (1913).

Compton's paper was:
A.H. Compton, *Phys. Rev.* **21**, 483 (1923).
The electron diffraction experiments of Davisson and Germer were published in:

C. Davisson and L.H. Germer, *Phys. Rev.* **30**, 705 (1927).

de Broglie's work is in English in:
L. de Broglie, *Phil. Mag.* **47**, 446 (1924).

Schrödinger's work
Useful accounts of his work (in English) are:

E. Schrödinger, *Phys. Rev.* **28**, 1049 (1926).
—, *Collected Papers on Wave Mechanics*, Blackie, London and Glasgow, 1928.

Statistical interpretation
Born's original work was published in:
M. Born, *Zeitschrift für Physik* **37**, 863 (1926).
—, *Zeitschrift für Physik* **38**, 803 (1926).

Also, see
—, *Physics in My Generation*, Springer, New York, 1969.

Taylor's experiments
G.I. Taylor, *Proc. Cambr. Phil. Soc.* **15**, 114 (1909).

Index

In the index page numbers may be followed by the letters m, q, or t. These refer to mathematical boxes, questions, and tables in the text respectively.